Autonomous Transport

Autonomous Transport presents contributions from the inaugural Autonomous Transport Conference held in Gdynia, Poland, in 2024 under Gdynia Maritime University (GMU). This comprehensive volume captures cutting-edge research and innovative insights from the scientific community on autonomous transport technologies across applications.

This book delivers state-of-the-art research on autonomous transport vehicle operations and new technology deployment. It addresses critical operational challenges, including the evolving requirements for maritime autonomous surface ships (MASS) navigation and necessary adjustments to established standards like the SOLAS convention and IMO instruments. Readers gain access to diverse perspectives on construction, operation, and the implementation of remote control and autonomous systems across multiple transport modes.

This book serves researchers, engineers, and professionals working in autonomous transport development, maritime operations, aviation technology, and land-based autonomous systems. It is particularly valuable for academics studying transport automation, industry practitioners implementing autonomous technologies, regulatory specialists adapting safety standards, and policymakers shaping the future of autonomous transport across all sectors.

Autonomous Transport

Edited by Teresa Abramowicz-Gerigk,
Joanna Soszyńska-Budny,
Zbigniew Burciu, and Adam Weintrit

CRC Press
Taylor & Francis Group
Boca Raton London New York

CRC Press is an imprint of the
Taylor & Francis Group, an **informa** business

Designed cover image: Shutterstock

First edition published 2026
by CRC Press
4 Park Square, Milton Park, Abingdon, Oxon, OX14 4RN

and by CRC Press
2385 NW Executive Center Drive, Suite 320, Boca Raton FL 33431

CRC Press is an imprint of the Taylor & Francis Group, an informa business

British Library Cataloguing-in-Publication Data
A catalogue record for this book is available from the British Library

ISBN: 9781041166832 (hbk)
ISBN: 9781041167228 (pbk)
ISBN: 9781003686088 (ebk)

DOI: 10.1201/9781003686088

Typeset in Times New Roman
by codeMantra

Contents

Preface

The book *Autonomous Transport* is a collection of selected articles presented at the 1st Autonomous Transport Conference, which took place on December 10, 2024, at the Faculty of Navigation of the Gdynia Maritime University. The conference, which will be held regularly in the future, aims to present research topics common to maritime, rail, road, and air autonomous means of transport. The conference is intended to be a forum for the exchange of views, research results, and implementations of new technologies in the field of autonomous transport.

The initiator of the 1st Autonomous Transport Conference is the Gdynia Maritime University in cooperation with the Motor Transport Institute in Warsaw and the Office of Rail Transport in Warsaw. The conference included all branches of transport, while selected articles concern autonomous maritime transport only.

The increase in the level of automation in the operation of transport units, including ships, together with the anticipated increase in the use of remote control, will require a new operating philosophy and a new approach to MASS (maritime autonomous surface ship) navigation. MASS will be inextricably linked to multimodal and intermodal freight transport systems. The collection of selected articles reflects these issues.

Editor biographies

Teresa Abramowicz-Gerigk is a Professor and Chair of the Department of Ship Operation at the Faculty of Navigation, Gdynia Maritime University. She gives lectures on applied mechanics and ship handling. Her field of interest is mathematical modeling of the safety of maritime transport and ship hydrodynamics. She has participated in several international scientific projects. She has published one book and over 100 papers in scientific journals and conference proceedings.

Joanna Soszyńska-Budny is a Professor at the Department of Transport at the Faculty of Navigation, Gdynia Maritime University. She gives lectures on mathematics, statistics, and reliability. Her field of interest is mathematical modeling of the safety and reliability of complex systems under variable operating conditions. She has participated in several international scientific projects. She has published three books and over 100 papers in scientific journals and conference proceedings.

Zbigniew Burciu, Master Mariner, specializes in maritime safety, transport safety, and maritime search and rescue. He is the author/co-author of over 180 publications and three monographs; manager of 12 research projects; promoter of four doctoral theses; and reviewer of four doctoral theses, four habilitation thesis, and four professorial theses. He has reviewed the award of honorary doctorates at the Maritime University of Gdynia.

Adam Weintrit, PhD, DSc, FRIN, FNI, Master Mariner, is the President of Gdynia Maritime University. His research interests encompass areas such as maritime navigation, maritime transport, geodesy, cartography, and maritime safety. He is the author of more than 350 scientific publications, including 18 books, and the editor of 19 monographs. The majority of his work is published in English. He held the position of Chair of the International Association of Maritime Universities (IAMU) from 2023 to 2025 and currently serves as an expert to the International Maritime Organisation (IMO). Since 2016, he has been a member of the Transport Committee of the Polish Academy of Sciences and was Chair of the Waterborne Section from 2016 to 2024. In 2024, he was re-elected to the Transport Committee and made a member of the Geodesy Committee of the Polish Academy of Sciences. He is the organizer of the 'TransNav' International Conference on Marine Navigation and Safety of Sea Transportation.

Contributors

Teresa Abramowicz-Gerigk
Department of Ship Operation, Faculty of Navigation
Gdynia Maritime University
Gdynia, Poland

Zbigniew Burciu
Department of Ship Operation, Faculty of Navigation
Gdynia Maritime University
Gdynia, Poland

Maciej Gogaliński
Faculty of Navigation
Maritime University of Szczecin
Szczecin, Poland

Maciej Gucma
Faculty of Navigation
Maritime University of Szczecin
Szczecin, Poland

Paulina Hatłas-Sowińska
Faculty of Navigation
Maritime University of Szczecin
Szczecin, Poland

Przemysław Jabłoński
Faculty of Navigation
Maritime University of Szczecin
Szczecin, Poland

Wojciech Koznowski
Faculty of Electrical Engineering
Gdynia Maritime University
Gdynia, Poland

Agnieszka Lazarowska
Faculty of Computer Science
Gdynia Maritime University
Gdynia, Poland

Józef Andrzej Lisowski
Faculty of Computer Science
Gdynia Maritime University
Gdynia, Poland

Andrzej Łebkowski
Faculty of Electrical Engineering
Gdynia Maritime University
Gdynia, Poland

Alicja Mrozowska
Faculty of Command and Naval Operations
Naval Academy
Gdynia, Poland

Piotr Mrozowski
Faculty of Navigation
Gdynia Maritime University
Gdynia, Poland

Monika Rybczak
Faculty of Computer Science
Gdynia Maritime University
Gdynia, Poland

Mirosław Siergiejczyk
Faculty of Transport
Warsaw University of Technology
Warsaw, Poland

Joanna Soszyńska-Budny
Department of Transport, Faculty of Navigation
Gdynia Maritime University,
Gdynia, Poland

Joanna Szłapczyńska
Faculty of Electronics, Telecommunications and Informatics
Gdansk University of Technology
Gdansk, Poland

Adam Weintrit
Department of Navigation, Faculty of Navigation
Gdynia Maritime University
Gdynia, Poland

Jakub Wnorowski
Faculty of Electrical Engineering
Gdynia Maritime University
Gdynia, Poland

Natalia Ziółkowska
Faculty of Electronics, Telecommunications and Informatics
Gdansk University of Technology
Gdansk, Poland

Chapter 1

Process failure and effects analysis of a sea voyage of MASS

Teresa Abramowicz-Gerigk, Joanna Soszyńska-Budny, Zbigniew Burciu, and Adam Weintrit

1.1 Introduction

Situational awareness (SA) of the remotely managed unit's journey by the operator will help prevent accidental events that disrupt operations. Full information will enable the trip to be maintained at the required level of safety. A high level of SA will allow the operator to counteract threats and predict their status in the near future in the technical and commercial operation of the ship. To facilitate the process of the regulatory scoping exercise, the degrees of autonomy are organized as follows [1]:

- **Degree one**: Ship with automated processes and decision support: Seafarers are on board to operate and control shipboard systems and functions. Some operations may be automated and at times be unsupervised, but with seafarers on board ready to take control.
- **Degree two**: Remotely controlled ship with seafarers on board: The ship is controlled and operated from another location. Seafarers are available on board to take control and to manage the shipboard systems and functions.
- **Degree three**: Remotely controlled ship without seafarers on board: The ship is controlled and operated from another location. There are no seafarers on board.
- **Degree four**: Fully autonomous ship: The operating system of the ship can make decisions and determine actions by itself.

The research [2] showed that despite the navigation equipment, sailors stayed on the navigation bridge; however, the most discussed keyword was "feelings" that they perceived by looking outside of the window and experiencing "standing wave", "rolling", or "sense of balance" with the vessel, which reduced their stress.

Therefore, unlike the ship's captain (degree one), who is on the navigation bridge, the operator in the remote operation center will create new elements in SA due to the lack of direct feel of the ship.

The risk in maritime autonomous surface ships (MASS) voyage planning is the probability of occurrence of a specific dangerous event (which may cause damage) combined with the consequences associated with this event (PN-IEC 300-3-9).

The risk assessment should be carried out systematically by e-Master supervised by the remote operation centre (ROC) operator in order to determine what potential hazards may occur in connection with the sea voyage, which may constitute a potential failure or maritime accident, and at the same time, whether they can be reduced and eliminated, and what actions should be taken.

DOI: 10.1201/9781003686088-1

1.2 Elements of situational awareness in shipping

SA increases safety, improves task performance, and enables correct decisions in the face of information overload.

SA in navigation will be the perception of elements related to the sea voyage, the technical and operational condition of the maritime unit, navigation and hydrometeorological conditions, and their understanding and prediction.

Carrying out a sea voyage will require considering 165 elements. The information was clustered into nine different groups [3].

1 Voyage
2 Sailing
3 Observations
4 Safety and emergencies
5 Security
6 Cargo, stability, and strength
7 Technical
8 Shore control center
9 Administrative.

1.2.1 *Process Failure Mode and Effects Analysis of a sea voyage of MASS*

From the areas presented in paragraphs 2–9, you can identify the risks involved, the possible errors of the operator during the implementation of the trip, and their potential consequences. The causes of these errors can also be identified, and an analysis of possible errors can be carried out, from which conclusions should be drawn to reduce their occurrence.

Below, in Figure 1.1, an analysis of the types of errors and their effects in the process of implementing a sea voyage for MASS degree I is presented. A MASS degree III is defined as a remotely operated vessel without seafarers on board. The ship is controlled and operated from another location.

With MASS III, the operator will not feel the so-called "standing wave," "rocking," or "sense of balance" affecting the vessel as a result of external conditions. This state of affairs can cause stress to the operator [2].

The assignment of risk indicators and the evaluation of each according to three criteria: severity (severity), probability of occurrence (occurrence), and possibility of detection (detection) allows for the determination of a risk priority index (Table 1.2).

The risk priority number (RPN) is the product of values Z, R, and W:

$$\mathrm{RPN} = Z \times R \times W \tag{1.1}$$

The indicator (RPN) allows to determine which failures require an immediate response.

Table 1.2 summarizes the priority number of risks.

List of possible errors during the sea voyage:

1 7 (Z) x 2 (R) x 3(W) = 42
2 **8 (Z) x 8 (R) x 3(W) = 192 (wrong ship's domain)**
3 9 (Z) x 2 (R) x 1(W) = 18
4 7 (Z) x 1 (R) x 1(W) = 7

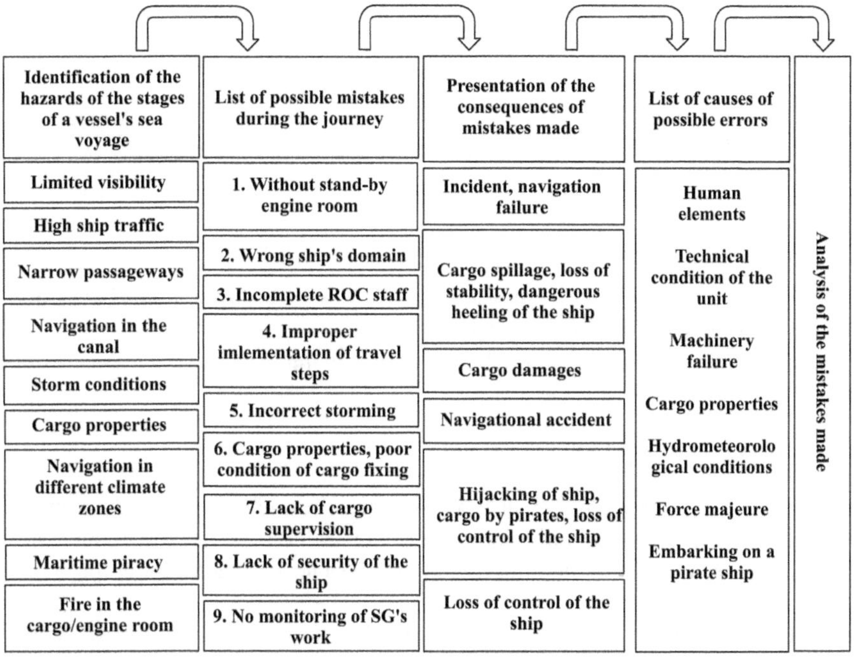

Figure 1.1 Application of the Process Failure Mode and Effects Analysis to identify MASS operator errors.

Table 1.1 Evaluation of MASS failures according to the three criteria

Significance for the safety of the realized journey (Z)		*Probability of occurrence/risk (R)*		*Detectability (W)*	
No perceived risk of failure	1	Highly unlikely	1	High possibility of error detection	1
Low risk of failure	2–3	Unlikely	2–3	Moderate possibility of error detection	2–3
Moderate risk of failure	4–6	Possible	4–6	Little chance of error detection	4–6
High risk of failure	7–8	Probable	7–8	Very little chance of error detection	7–8
Very high risk of failure	9–10	Certain	9–10	Unlikely error detection	9–10

Table. 1.2 Risk priority number

Type of risk	*RPN values*
Low risk	1–49
Medium risk	50–99
High risk	100–199
Very high risk	200–1000

5 9 (*Z*) x 2 (*R*) x 1(*W*)=18
6 9 (*Z*) x 1 (*R*) x 1(*W*)=9
7 9 (*Z*) x 2 (*R*) x 2(*W*)=36
8 4 (*Z*) x 2 (*R*) x 1(*W*)=8
9 9 (*Z*) x 1 (*R*) x 1(*W*)=9

From the conducted Process Failure Mode and Effects Analysis (PFMEA), it is concluded that the existing risks of the realized sea voyage may occur in the following areas:

- human factor,
- technical condition of the unit,
- cargo,
- hydrometeorological conditions, and
- external causes.

The result of the analysis showed that the risk of the wrong ship domain being used by the operator, (*Z*) x 8 (*R*) x 3(*W*) = 192, is very high.

The consequences of these errors and the impact of the human factor can range from incidental to catastrophic events.

1.2.2 Human elements

In [4], the role of the human factor in navigation was presented. It also presented how the development of autonomous technologies will change seafarers' skills and competencies in the future. It was pointed out that in the future, there will be a need to develop the following factors:

- Increased knowledge on communications and information technology for both shore-based and ship-based personel.
- Understanding of technology associated with Marine Autonomous Surface Ships (MASS), its possibilities, and its limitations.

MSC 108 had agreed to develop high-level training provisions for the MASS Code, where competency Knowledge, Understanding, and Proficiencies (KUP) can be developed by IMO's Human Element, Training and Watchkeeping Sub-Committee (HTW) following the finalization of the Code. At this stage, the MASS Code is still to be completed, so it will be considered later.

A discussion around having a sole watchkeeper on MASS at night for trials took place. HTW will advise the MSC that International Convention on Standards of Training, Certification and Watchkeeping (STCW) regulation 1/13 (conduct of trials) is applicable to MASS trials. This will allow the MSC to decide if a sole watchkeeper at night is permitted.

1.2.3 Ship domain

From the International Dictionary of Marine Aids to Navigation. VTS-2012–30 ship domain is defined as follows:

- An operational zone around, above, or below a vessel within which an incursion by another fixed or moving object, or another domain, may trigger reactions or processes.
  or
- The area around the ship that should be clear of other vessels or objects—depends on many factors [5].
  or
- A ship domain may be thought of as the sea around the ship, which the navigator would like to keep free with respect to other ships and fixed objects [6].
  or

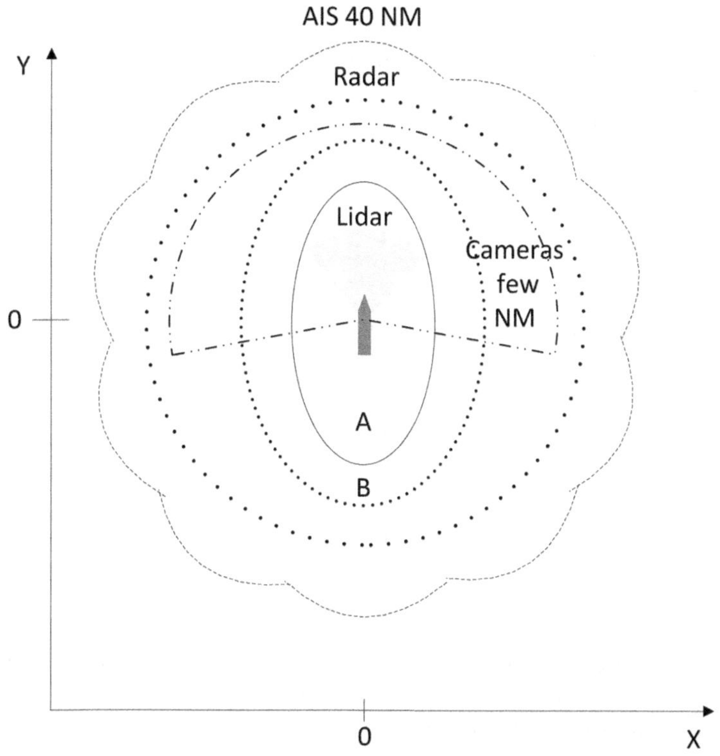

Figure 1.2 Ship domain: (a) manned ship domain and (b) remotely controlled MASS domain.

- The concept of a safe navigational domain around the ship was created to determine the safety of ship traffic in restricted areas. The domain can be defined as the area around the ship, dependent on vessel parameters and traffic situation, in which no other objects should be present [7].

An extension of the concept of the ship's domain is the fuzzy ship's domain [8], defined as the area around the ship that the ship's navigator should keep free from other vessels and objects, the shape and size of which depend on the adopted level of navigational safety, understood as the degree to which the navigational situation belongs to the fuzzy set "dangerous navigation" [8].

The size of the domain will depend on the size of the ship, the maneuvering and operating characteristics, and the characteristics of the navigation equipment.

The example of ship domains determined by the Ship Master and e-Master as safety zones is presented in Figure 1.2. The figure shows the approximate boundaries of the ship safety navigational area from their perspective [9,10].

1.3 MASS operator attitudes toward risk

The MASS III operator will take several actions to ensure a safe journey and should be aware that the decisions he or she makes depend on psychological factors, professional factors, and those supporting his or her work in the ROC. The decisions he or she makes under risk will ultimately be subjective because they depend on his or her perspective and opinion. In view of the above, a subjective approach to risk analysis is introduced by utility theory, applying the principle of maximizing expected utility [11].

In making a decision under risk, the operator can be guided by the criterion of the expected value of the utility function, which is a weighted average of the utility values of all possible outcomes, with the weights equal to the probabilities of their occurrence.

In mathematical terms, we express this with the following formula:

$$E[U]=\sum_{i=1}^{n} p_i U\left(x_i\right), \tag{1.2}$$

where

$E[U]$—expected utility value,

p_i—probability of occurrence of the i-th result (where $\sum_{i=1}^{n} p_i = 1$),

$U\left(x_i\right)$—the value of the utility function related to the result x_i.

The value of the utility function $U\left(x_i\right)$ depends on the risk approach a given MASS III remote operator has. The way an autonomous unit operator approaches risk affects the shape of its utility function.

The utility function expresses satisfaction or psychological comfort. It can be shown that between the form of the utility function and the attitude of the individual operator to risk exist the following relationships:

- An operator with a concave utility function is characterized by risk aversion—aversion to making decisions with a high probability of risk, a tendency to reduce risk, such a person will prefer specific outcomes over risky ones;
- An operator with a convex utility function is characterized by a risk-taking propensity—making decisions with a high probability of risk, such a person will prefer options with more risk, even though they may involve uncertainty;
- An operator with a linear utility function is characterized by indifference to risk.

These relationships are shown in Figure 1.3.

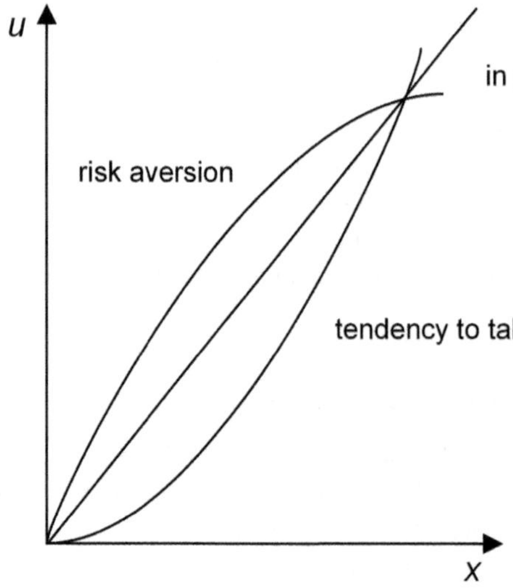

Figure 1.3 The shape of the utility function in the context of the risk approach.

Taking into account the shape of the utility function, we can define its general form as follows:

$$U(x) = x^{\alpha} \tag{1.3}$$

where α is a parameter that determines the MASS unit operator's approach to risk. Thus,

- if $\alpha < 1$—the person shows risk aversion (concave function),
- if $\alpha = 1$—the person is neutral to risk neutral (linear function),
- if $\alpha > 1$—the person has a propensity for risk (convex function).

Lack of "feeling" of the marine unit—MASS by the operator will cause stress. It should be noted that prolonged stress, and consequently persistent symptoms, increases the risk of error over time. Meanwhile, stress is one of the most common risks in today's work environment [12] (Figure 1.4).

Stress can cause a MASS operator's attitudes toward risk to be more "pronounced" (Figure 1.5), which can influence his decision-making, such as decisions about the size of a ship's domain.

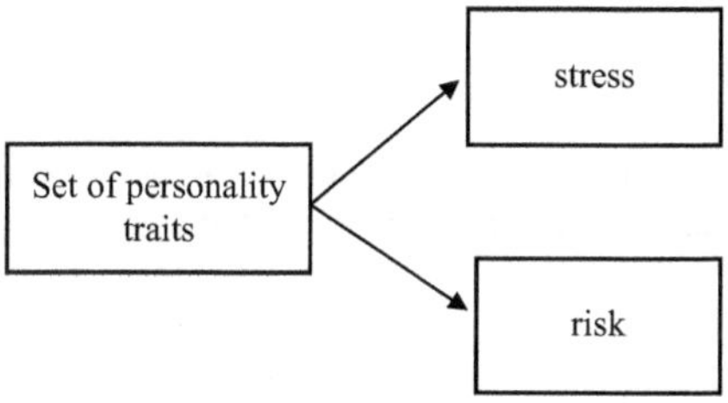

Figure 1.4 The impact of temperament on risk and stress [12].

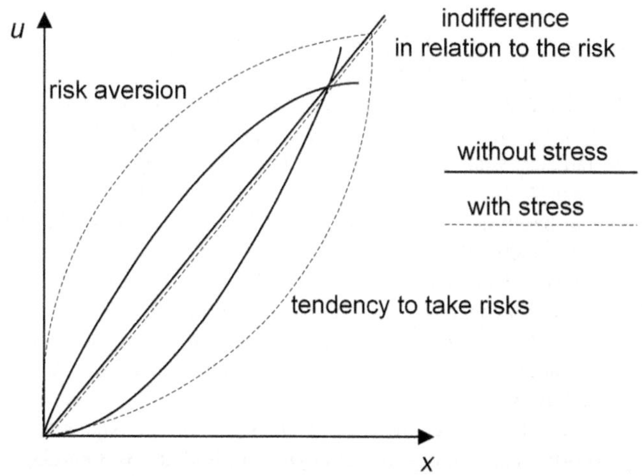

Figure 1.5 The shape of the utility function in the context of the risk approach. Types of operators' approach to the executed sea voyage with consideration of MASS III operator stress.

1.4 Conclusion

Presenting PFMEA enables the assessment and identification of potential threats that may occur during a MASS sea voyage and allows analysis of possible errors.

In the literature, we find studies identifying the relationship between performance effectiveness and the personality and temperament of pilots [13,14]. This research has contributed to reducing the number of pilot errors.

The MASS III operator will take several actions to realize a safe voyage. When making decisions, he or she should be aware of the unit's lack of "feel". It should also be noted that the decisions made will depend on psychological factors, professional factors, and ROC facilitators.

The operator's lack of "feel" for MASS III may cause changes in the operator's attitude toward the risks of the sea voyage being pursued. Chronic stress also contributes to a reduction in overall operator productivity. The use of autonomous units on waterways may result in the threat of a new quality of hazards not currently present in maritime transportation, which will occur with low probability but high consequences.

According to the PFMEA of a sea voyage of MASS, the assumed incorrect domain of the ship causes a very high risk of failure. To reduce risks, the usage of AI assistants in ROCs, i.e., artificial intelligence systems, should be considered to support the decision-making process of the MASS operator.

References

1. MSC 100/20/Add.1 Annex 2, page 1. Annex 2 Framework for the regulatory scoping exercise for the use of maritime autonomous surface ships (MASS). https://maiif.org/wp-content/uploads/2019/06/MSC-100_20-Annex-20-1.pdf
2. Man, Y., Lundh, M., Porathe, T. Seeking harmony in shore-based unmanned ship handling - from the perspective of human factors, what is the difference we need to focus on from being onboard to onshore? *Proceedings of the 5th International Conference on Applied Human Factors and Ergonomics AHFE 2014*, Kraków, Poland 19–23 July 2014. https://doi.org/10.54941/ahfe100625
3. Porathe, T., Prison, J., Man, Y. Remote control centres for unmanned ships. *Chalmer University of Technology*, Sweden. *Human Factors in Ship Design & Operation*, 26–27 February 2014, London, UK. © 2014.
4. IMO Human Element, Training and Watchkeeping (HTW 11). *Summary Report Lloyd's Register Briefing Note – Summary Report HTW 11*, 2025.
5. Pietrzykowski, Z., Wielgosz, M., Siemianowicz, M. In the restricted area – simulation research. *Scientific Journals Maritime University of Szczecin*, 32(104), 152–156, 2012.
6. Goodwin, E. M. A statistical study of ship domain. *Journal of Navigation*, 28, 329–341, 1975.
7. Hansen, M. G., Jensen, T. K., Lehn-Schiøler, T., Melchild, K., Rasmussen, F. M., Ennemark, F. Empirical ship domain based on AIS data. *Journal of Navigation* 66(6), 931–940, 2013. https://doi.org/10.1017/s0373463313000489
8. Pietrzykowski, Z., Maga, J. Domena statku w identyfikacji incydentów w systemach rozgraniczenia ruchu. *Prace Naukowe Politechniki Warszawskiej z. Transport* 114, 279–290, 2016.
9. Abramowicz-Gerigk, T., Burciu, Z. Situational awareness in autonomous shipping – ship domain in remote MASS operation. *The International Journal on Marine Navigation and Safety of Sea Transportation* 18(4), 889–893, 2024.
10. Thombre, S., Zhao, Z., Ramm-Schmidt, H., García, J. M. V., Malkamäki, T., Nikolskiy, S., Hammarberg, T., Nuortie, H., Zahidul, M., Bhuiyan, H., Särkkä, S., Lehtola, V.V. Sensors and AI techniques for situational awareness in autonomous ships: a review. *IEEE Transactions on Intelligent Transportation Systems*, 23(1), 64–83, 2022.

11. Jajuga, K. *Zarządzanie ryzykiem*. Wydawnictwo Naukowe PWN, Warszawa, 2007.
12. Makarowski, R. The human factor in aviation. *Przegląd psychologiczny* 55(3), 305–326, 2012.
13. Makarowski, R., Piotrowski, A. Profil psychologiczny pilotów samolotów pasażerskich–analiza temperamentu, agresji i skłonności do podejmowania ryzyka. *Medycyna Pracy*, 68(5), 639–651, 2017. https://medpr.imp.lodz.pl/pdf-68932-12683?filename=Profil%20psychologiczny.pdf
14. Mitchell, H. D. Aircrew selection. *American Journal of Psychiatry* 99, 354–357, 1942. https://doi.org/10.1176/ajp.99.3.354

Chapter 2

MANET networks as a key system for autonomous navigation advancement

Maciej Gucma, Paulina Hatłas-Sowińska, Przemysław Jabłoński, and Maciej Gogaliński

2.1 Introduction

One of the key issues in any digital data exchange system, especially in semi- or fully autonomous vessels, is communication between nodes (terminals or sensors). These end devices (simplified as vessels) initially communicate in one-to-one or one-to-all mode via standard radio equipment described by GMDSS (global maritime distress and safety system) in the maritime world. Such communication is insufficient for independent data transmission between semi- or fully autonomous vessels. The potential of mobile ad hoc networks (MANETs) to provide flexible and robust communication solutions makes them an appealing choice for maritime autonomous systems operating in areas without existing infrastructure. Since the MANET concept is not new, its use in MAAS (Maritime Autonomous Surface Ships) is an emerging application. Ad hoc networks can effectively support multiple maritime functions, including autonomous ship navigation systems and communication networks for offshore applications. Integrating MANETs into maritime systems introduces exclusive challenges and opportunities that researchers and developers are presently investigating.

2.2 Related work

Recent research shows that the subject has been investigated in the scope of autonomous ship guidance problems. MANETs are increasingly being explored in maritime autonomous systems because they can provide flexible, cost-effective communication solutions in environments where traditional infrastructure is limited or costly. These networks can support a range of maritime applications, from autonomous ship guidance to data communication for fishing vessels. Integrating MANETs in maritime systems offers several advantages, including enhanced connectivity, scalability, and adaptability to dynamic maritime conditions. Below, key aspects of MANETs in maritime autonomous systems are discussed and, for clarity, divided into sections.

2.2.1 Design and implementation of maritime MANETs

A MANET designed under the guidance of an autonomous ship can effectively manage multi-modal vessel interactions. The proposed system includes network composition, mode, and management methods, ensuring robust connectivity and transmission capacity among ships [1]. MANETs offer a viable alternative to expensive satellite communications for maritime vessels. By dynamically adjusting transmission power levels and routing options, MANETs can achieve satisfactory throughput and packet error rates, making them suitable for fishing vessels and other maritime applications [2].

DOI: 10.1201/9781003686088-2

2.2.2 *Routing protocols and network performance*

The Normalized Location and Transmission Characteristics Routing (NLTCR) scheme optimizes routes based on distances to destination ships and transmission characteristics, enhancing connectivity and scalability in maritime environments [3]. The AIS-aided ad hoc on-demand distance vector (AODV) (A-AODV) protocol reduces routing overhead by utilizing ship location information and supporting multi-hop broadband data transmission with lower overhead [4]. This chapter introduces an improved AODV routing protocol designed explicitly for MANETs, facilitating efficient ship-to-ship and shore-to-ship wireless broadband communications. Developing maritime MANETs using 27/40 MHz bands has demonstrated effective multi-hop communication and transmission power control, which is crucial for maintaining network performance in maritime settings [5]. This research focuses on implementing an AODV routing protocol with transmission power control to improve communication range and efficiency.

2.2.3 *Challenges and opportunities*

Maritime networks face unique challenges due to their combination of mobile nodes and fixed satellite links. It highlights the need for efficient management solutions that can adapt to the unique operational requirements of maritime environments, emphasizing the importance of automation and resource optimization. The authors propose a policy-based service-oriented architecture to facilitate rapid deployment of management services and to improve traffic engineering through traffic monitoring, prioritization, and adaptive routing techniques. Efficient network management is critical, with opportunities for automation and resource optimization to improve service delivery and traffic engineering [6]. Maritime traffic patterns, node densities, and mobility behaviors influence the performance of MANET routing protocols. Protocols like AODV, AOMDV (ad-hoc on-demand multipath distance vector), and DSDV (destination-sequenced distance-vector) have been evaluated for their effectiveness in such environments, with varying results based on specific conditions [7].

2.2.4 *Technological innovations*

Another proposed system uses multiple directive beams for transmission, offering longer transmission ranges and uniting high capacity, flexibility, and self-organizing capabilities, which are essential for maritime communication networks [8]. It highlights the development of a distributed surveillance application called "Sensors in Concert" (SinC), which leverages maritime MANET for continuous and complete surveillance of surface objects in large maritime areas. The paper [9] discusses the unique challenges of mobile ad hoc networking in maritime contexts and outlines the architecture and components that address these challenges. The cited paper focuses on a specific application of MANET called vehicular ad hoc networks (VANET), which typically deals with vehicles on roads. However, this research explores the idea of applying similar technology to ships at sea. Modeling the networks is another critical issue that shall be followed up on, along with the search for optimal protocols. Authors [10] aim to implement the MANET concept alongside ontological-based communication at sea.

Another important advance in MANET development is tactical networks, where several breakthroughs were made last year. Since the environment and concept differ, we present this here only as a reference. The paper [11] addresses the challenges of tactical edge networks, which are characterized by fluctuating link quality and changing topologies, often lacking pre-deployed infrastructure, leading to the use of MANETs in these environments. The paper

[12] presents a novel methodology that utilizes stochastic distributions to estimate control phase time (CPT) for software-defined radios (SDRs) in tactical MANETs. This approach aims to enhance network connectivity by evaluating the lifetime of communication links based on the distances between radios and their transmission ranges, ultimately improving the reliability and efficiency of control messaging in various mobility scenarios.

2.3 MAAS application

Applying autonomous or semi-autonomous vessels in merchant shipping requires a series of technological, economic, legal, and societal breakthroughs. The application requires delving into the following issues:

- Integration with existing infrastructure
- Regulatory and legal uncertainty
- Cybersecurity
- Reliability and safety of subsystems
- Terrorism and piracy
- Economic constraints
- AI control and transparency.

Since the technology exists and is field-proven in other fields, communication seems to be a solved problem. Its position in the maritime world is significantly delayed in terms of local solutions. Namely, most of the vessl2vessel communication lies on VHF (very high-frequency) radio voice transmissions, with limited capabilities of channel capacity. AIS (automatic ship identification system) exchanges only crucial navigation information; its delay and small bandwidth do not look promising for higher throughputs. There are no communication systems capable of exchanging information between vessels efficiently and openly. Vessel operators rely on satellite systems (like the widely used Starlink or Iridium). This type of communication requires the global WAN (wide area network) to operate in the discussed applications. Several types of radio communications have been created for tactical and land-based SAR (search and rescue) applications. They rely on MANETs and radios that use advanced ways of beamforming, along with the SDRs, where the waveform is created by software and can be changed during operation. Core concepts that shall be investigated in the MANET-type vessel communication for MASS operation are:

- Routing protocol
- Beamforming and antenna array
- Bandwidth related to distance
- Safety of network
- Redundancy.

2.4 Construction of MANET network

The primary idea of MANET networking is to communicate between points that are not visible (in radio terms) to each other. In Figure 2.1, vessels are represented as network nodes, where #1 and #4 are not in direct radio coverage. Communication can still be achieved through primary MANET routing via nodes #2 and #3 or through redundant satellite links. Specialized routing

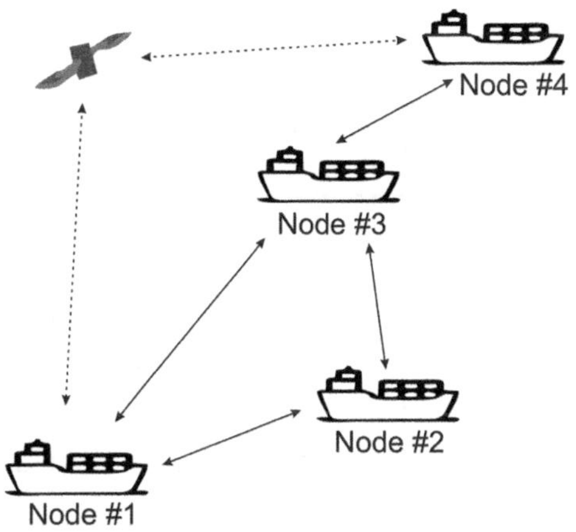

Figure 2.1 Communication between nodes in MANET.

protocols facilitate efficient communication in MANETs in environments without infrastructure that exhibit dynamic characteristics. These protocols are broadly classified into three categories: Proactive routing operates differently from reactive and hybrid protocols because each category possesses unique operational philosophies and key features. Table-driven protocols, called proactive protocols, keep routing tables up to date across all network nodes through regular control message exchanges and updates. This method guarantees immediate accessibility to all destinations' routes, ensuring low data transmission latency [13]. The ongoing exchange of routing data results in significant overhead and power consumption, particularly in dynamic or extensive network structures. The DSDV and OLSR (optimized link state routing) protocols are examples of proactive routing protocols [14].

On-demand protocols establish network routes only when active data transmission is required. The source node triggers a route-discovery process to establish the best path to reach the destination whenever communication is desired. By keeping routes active solely for current connections, this method reduces control overhead and conserves bandwidth usage [13]. The primary downside of this system is the extended latency during route establishment, which affects applications that require immediate data delivery. AODV, DSR, and TORA (temporally ordered routing algorithm) represent examples of reactive routing protocols. Hybrid protocols combine proactive and reactive methodologies to effectively manage operational overhead and responsiveness. Such protocols preserve active routing data within a local area but use reactive route-discovery methods when accessing destinations outside this area [15]. The design demonstrates high effectiveness in large, diverse network environments by reducing control traffic and providing fast access to popular routes. ZRP (zone routing protocol) and SHARP (sharp hybrid adaptive routing protocol) are well-known instances of hybrid routing protocols. MANET routing protocol selection depends on network-specific requirements, including size, node mobility, and application needs. Proactive protocols enable low-latency communication but incur higher overhead costs. In contrast, reactive protocols reduce overhead but introduce route-discovery delays, and hybrid protocols aim to balance both dimensions to achieve scalable and efficient network performance. Figure 2.2 presents types and divisions of routing protocols. The description of all the mentioned routing protocols' advantages and weak sides is beyond the scope of this chapter but can be found in many works.

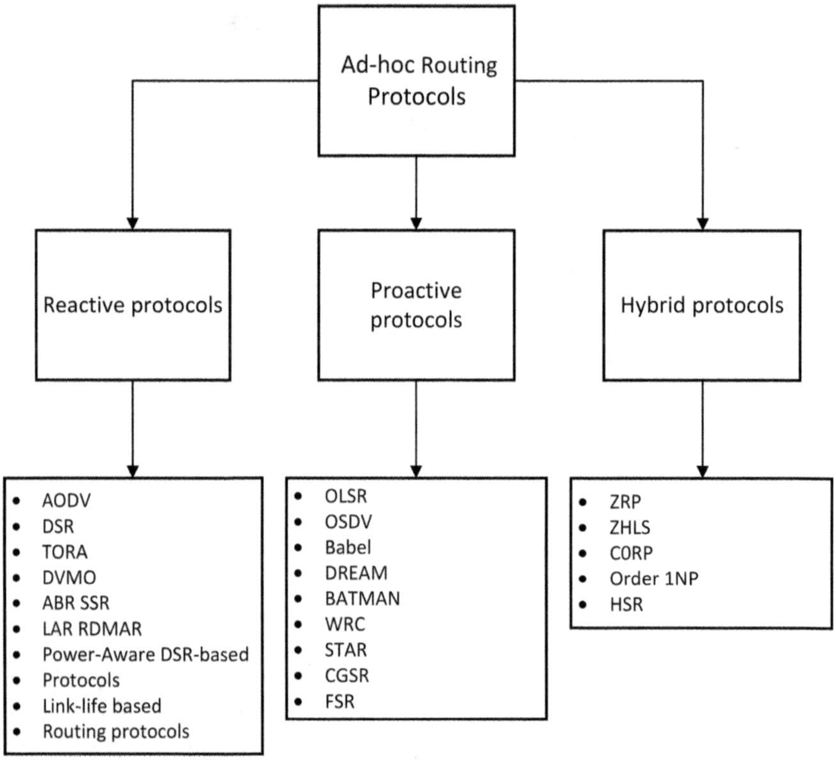

Figure 2.2 Division of routing protocols.

There are also geographical types of MANET protocols based solely on connections established by comparing the geolocation of nodes. These are more like the hybrid protocols with some additional features. It should be noted that geographical protocols can emerge, as they did in the car industry. The choice of routing protocol depends on the MAAS application, which relies on some basic assumptions:

1 Lack of coordinator—every node can route other packets without supervision;
2 Acknowledge that the recipient received information that can travel another route than the original message was received;
3 Transmission latency is higher as distance rises and shall be minimized on smaller ones;
4 Mission-critical (like the safety of navigation) transmissions are prioritized.

Another critical factor is the bandwidth of the communication channels. Since the typical AIS or DSC (digital selective calling) is pretty low (around 1200 baud for DSC), the choice of channel and type of modulation for future maritime-based MANET is critical here. Omni-direction antennas provide coverage for all vessels in the vicinity, although they require a lot of power and have low efficiency. Results for such tests with an omnidirectional antenna for the throughput of the network have been presented in the work [16]. Directional antennas with high gain can be utilized here. Since this sounds great, a solution would require stabilizing the antenna or utilizing narrow horizontal and wide vertical beams for the marine environment. Using a modified inverse power-law relationship, which offers a simplified construction of the relation between distance and throughput, we get [16]:

$$T(d) = \frac{T_{max}}{1 + \left(\frac{d}{d_0}\right)^{\eta}} \tag{2.1}$$

where

T_{max}: Maximum achievable throughput at $d = 0$

d_0: Reference distance where throughput begins to degrade significantly

η: Decay exponent modeling path loss and environmental effects (typically $2 \leq \eta \leq 52 \leq \eta \leq 5$ for wireless channels in 2.4 GHz) [17].

Equation (2.1) can be modified to obtain the specific environment and conditions of the network. For the realization of network parameters, modeling software shall be used, like CloudRF or Tetcos. We assumed the network for a small distance (up to 1 km) and generic radios with omnidirectional antennas. This setup uses ISM band 92.4 GHz and 870–900 MHz with radios working with power = 100 mW. Results of throughput in a generalized no-obstruction environment are presented in Figure 2.3.

We observe that ISM radio in a lower band with a similar pattern and modulation type achieves lower throughput than at higher frequencies, where no signal can be received. Further research

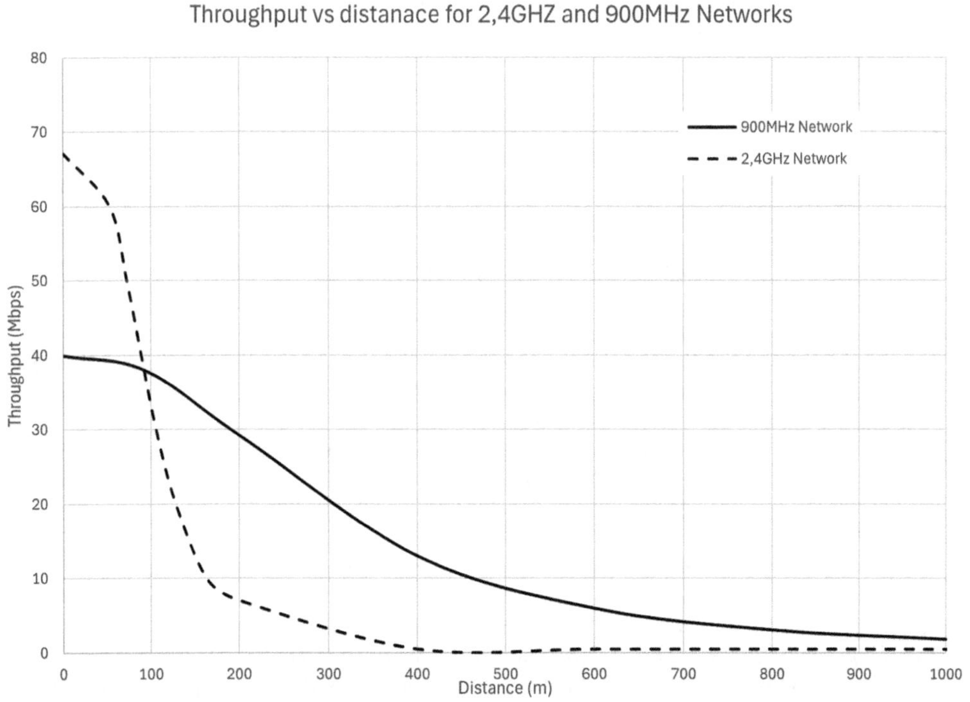

Figure 2.3 Comparison of throughput for 2.4 GHz and 870–900 MHz in distances 0–1000 m.

is in the field of directional high-gain antennas for 2.4 GHz, where its automatic direction setup is available with location messages from another node.

The safety of the network is another factor; here, it should be noted that modern fully open AIS transmissions are highly prone to cyberattacks. In several cases, AIS was jammed or spoofed by 3rd parties, resulting in a complete stoppage of helpful signal reception. Some of the cases have been presented in the articles provided by Slovenian researchers [18].

Redundancy can be achieved through different communication channels. Since most vessels are equipped with satellite transmission, it's the easiest way to join all the nodes to the WAN via this type of link. Here, the vessels will be treated as IoT (Internet of Things) end devices, and such work is ongoing [19].

2.5 Summary

This chapter explores the application of MANETs as a transformative solution for communication in Maritime Autonomous Surface Ships (MASS) and other autonomous maritime systems. It discusses the limitations of traditional maritime communication methods. It highlights the advantages of MANETs, including flexibility, scalability, and the ability to support robust, multi-hop wireless connectivity in infrastructure-less environments. The study reviews recent research on MANET design, implementation, and routing protocols—such as AODV, AOMDV, and DSDV—evaluating their performance under varying maritime conditions and traffic patterns. This chapter also addresses the challenges of maritime MANET deployment, including dynamic topology, variable node density, cybersecurity, and regulatory issues. Technological innovations such as beamforming, SDRs, and adaptive routing are examined for their potential to enhance throughput and reliability. This chapter concludes by emphasizing the need for further research in network management, protocol optimization, and integration with existing maritime MAAS transformation.

References

1. Wang, X., Zhang, Y., Liu, Z., Wang, S. and Zou, Y. 2023. Design of multi-modal ship mobile ad hoc network under the guidance of an autonomous ship. *Journal of Marine Science and Engineering* 11(5), 962. https://doi.org/10/gt3q6g
2. Gangaprasad, G. and Seetha Ramanjaneyulu, B. 2023. Ad-hoc marine networks suitable for the data communication needs of fishing vessels. *2023 14th International Conference on Computing Communication and Networking Technologies* (ICCCNT) (pp. 1–7). IEEE, New Delhi, India.
3. Son, J. 2013. A multi carrier selectable routing scheme by normalized transmission characteristics (MCS-NTC) at marine multi-carrier MANETs. *Journal of the Korean Society of Marine Engineering* 37(2), 199–204. https://doi.org/10.5916/jkosme.2013.37.2.199
4. Peng, S., Wang, Y., Xiao, H. and Lin, B. 2020. Implementation of an improved AODV routing protocol for maritime ad-hoc networks. *2020 13th International Congress on Image and Signal Processing, Biomedical Engineering and Informatics (CISP-BMEI)* (pp. 7–11). IEEE.
5. Yoshikawa, T., Kawasaki, S., Takase, M., Hiraoka, Y., Takaki, Y., Ohta, C. and Inoue, T. 2010. Development of 27MHz/40MHz bands maritime wireless ad-hoc networks. *2010 Second International Conference on Ubiquitous and Future Networks (ICUFN)* (pp. 177–182). Jeju Island, Korea.
6. Kidston, D. and Kunz, T. 2008. Challenges and opportunities in managing maritime networks. *IEEE Communications Magazine* 46(10), 162–168. https://doi.org/10.1109/MCOM.2008.4644135
7. Mohsin, R.J. and Woods, J. 2014. Performance evaluation of MANET routing protocols in a maritime environment. *2014 6th Computer Science and Electronic Engineering Conference (CEEC)* (pp. 1–5). Colchester, United Kingdom.

8. Laarhuis, J. H. 2010. MaritimeManet: mobile ad-hoc networking at sea. *2010 International WaterSide Security Conference* (pp. 1–6). IEEE, Marina di Carrara, Italy.
9. Pullin, A.J., Presland, S. and Pattinson, C. 2008. Using ship movement in the Irish Sea for MANET Evaluation. *European Symposium on Computer Modeling and Simulation* (pp. 394–399). https://doi.org/10.1109/EMS.2008.50
10. Wielgosz, M., Hatłas-Sowińska, P. and Gucma, M. 2025. Anticollision communication ontology technique implementation with MANET networking. *Scientific Reports* 15(1), 14416. https://doi.org/10.1038/s41598-025-98621-9
11. Suri, N., Campioni, L., Di Caro, E., Breedy, M., Morelli, A., Fronteddu, R. and Tortonesi, M. 2023. Adaptive information dissemination over tactical edge networks. *2023 International Conference on Military Communications and Information Systems (ICMCIS)* (pp. 1–7). IEEE: Skopje, North Macedonia.
12. Rukaiya, S.A.K., Farooq, M.U., Khattak, M.A., Matloob, I. and Nosheen, I. 2021. A stochastic distribution based methodology to estimate control phase time for software-defined radios in tactical MANETs. *IEEE Access* 9, 71687–71698. https://doi.org/10.1109/ACCESS.2021.3079173
13. Nwuku, Y.E., Ajibesin, A.A. and Ishaq, A.T. 2021. A comprehensive analysis of proactive and reactive manet routing protocols using Ns3. *Journal of Engineering and Applied Sciences Technology* 3(1), 1–6. https://doi.org/10.47363/JEAST/2021(3)120
14. Clausen, T. and Jacquet, P. 2003. *Optimized Link State Routing Protocol (OLSR)*. Technical Report #RFC3626. RFC Editor. https://www.rfc-editor.org/rfc/rfc3626.html
15. Biswas, A.K. and Dasgupta, M. 2020. A secure hybrid routing protocol for Mobile Ad-Hoc Networks (MANETs). *2020 11th International Conference on Computing, Communication and Networking Technologies (ICCCNT),* Kharagpur, India, July 2020.
16. Wang, W., Sharif, H., Hempel, M., Zhou, T. and Mahasukhon, P. 2008. Throughput vs. distance trade-offs and deployment considerations for a multi-hop IEEE 802.16e Railroad Test Bed. *VTC Spring 2008- IEEE Vehicular Technology Conference*, Marina Bay, Singapore, May 2008 (pp. 2596–2600).
17. Tetcos: NetSim – Network Simulation Software, India. 2021. https://www.tetcos.com/index.html
18. Androjna, A., Perkovič, M., Pavic, I. and Mišković, J. 2021. AIS data vulnerability indicated by a spoofing case-study. *Applied Sciences* 11(11), 5015. https://doi.org/10.3390/app11115015
19. Bouhlal, A., Abdelouahed, R.A., Marzak, A. and Meriem, B. 2023. Proposal to evaluate the integration of IoT technologies in the maritime domain. *Procedia Computer Science* 220, 1057–1064. https://doi.org/10.1016/j.procs.2023.03.148

Chapter 3

Possibilities of using an ontology system in autonomous states for communication purposes

Paulina Hatłas-Sowińska

3.1 Introduction

Autonomous surface vessels, particularly naval vessels, have been a hot topic for many years. The factors expected to drive the growth of autonomous ships include growing investment in autonomous ship ventures across many countries, the adoption of fully autonomous commercial and military vessels, and the increased use of automated systems to reduce human error [1]. Despite the global interest and development of this field, slower progress in this area is related, among other things, to the complex problems of safety and security of these means of transport. Especially in difficult-to-navigate areas with high traffic intensity, access to information on the current navigation situation, including the parameters of other ships' movements, is of great importance. A major obstacle may be both the lack of access to information necessary in a given situation, its excess, and the related difficulties in processing it, and the problems in communication between participants in the transport process, mainly between navigators steering ships. In the latter case, these include, among others, lack of communication, misunderstanding of the received message, incorrect selection of the message, or incorrect interpretation of the information exchanged. In connection with the above, both the automation of the processes of obtaining, selecting, processing, and presenting navigational information and the automation of communication processes carried out so far mainly by navigators steering ships and the personnel of land centers are becoming important.

In this chapter, the author focuses on the automation of the above-mentioned processes based on the author's own navigation ontology, communication ontology, and metaontology in the analyzed scope.

The goal to establish direct communication between ships and automate communication processes can reduce erroneous decisions and, as a result, erroneous actions that lead to maritime accidents.

3.2 Communication

Communication errors between navigators steering ships are one of the causes of maritime accidents. These include, among others, misunderstanding or lack of information. The article [2] presents the issues of identifying a navigational situation requiring communication and the tasks for the proposed system. It is well known that when defining communication areas, the division used in GMDSS (Global Maritime Distress and Safety System) is adopted. It specifies the rules and procedures for communication. However, it is often required to obtain additional

DOI: 10.1201/9781003686088-3

information or change decisions through communication by both navigators on ships and shore personnel. This involves additional information or obtaining data through dialogue.

The complexity of the decision-making situation may require not only obtaining additional information but also carrying out agreements between the parties. For example, they may concern planned own actions or expected actions of the other party, and their need most often results from a discrepancy in goals, methods of analysis, or assessment of the situation. An example may be communication to ensure safety in the event of an encounter between two ships, aimed at ascertaining the navigator's intentions on the other ship, making a decision on the safe way of passing the units, and determining maneuvers to be performed by both ships. The introduction of ontology—as the basis for communication—will allow for: unification of concepts, lack of unnecessary messages, unambiguity, and precision of the response. The most crucial point is the unambiguous interpretation of the message sent. The system is built so universally that it could operate not only in the navigator-navigator system but also M2M (machine to machine).

3.3 Ontology

By definition, an ontology is a theory of any field that describes concepts hierarchically to establish semantic relations. It is characterized by a logical theory that introduces restrictions on logical models. The basis for both understanding and generating a message is the appropriate interconnection of the meaning of concepts (words) occurring in the message. Due to the different requirements placed on information exchange and negotiation processes, it becomes necessary to distinguish specialized ontologies (here: communication and navigation). It should be remembered, however, that it is impossible to use them alone—it is necessary to link the communication ontology with a second ontology, e.g., navigation of a given transport field. In maritime navigation, standard phrases were mainly used for communication at sea [3].

Ontology deals with discovering and describing "what is"—in reality. It is what allows us to formalize knowledge and shows, in an unambiguous way, the representations of a given knowledge. It describes concepts hierarchically to establish semantic relations in a given domain. It is characterized by a logical theory that introduces constraints into logical models. It is defined, among others, as conceptualization [4].

An ontology can be written as a formula [5], where the set O defines the structure of concepts and the relations between them, as well as the theory of the model being described.

For any broad domain, we can create different ontologies to describe the domain in various ways, so this ontology has been described using a new, proprietary, detailed formula to facilitate the collection of words for maritime transport.

$$O = \{A, K, R, f, g\} \tag{3.1}$$

where
O: ontology
A: axiom of choice
K: class of abstraction
R: relationship
f: message-generating function
g: message interpretation function.

The basic rules of each element in the set O are as follows [6,7]:

The authors' original objective function that, in combination with the existing ontology, forms a metaontology applicable to automated communication. The objective function includes the message type and specific category and determines which connections are possible and which are rejected.

$$f_{O=m_t(A)\cdot\sum_{n=1}^{4}A_n+m_t(Q)\cdot\sum_{n=1}^{3}Q_n+m_t(T)\cdot\sum_{n=1}^{7}T_n} \quad (3.2)$$

where
f_O: objective function
m_t: message type

$m_t(A)$: answer
$m_t(Q)$: question
$m_t(T)$: tell

A_n: n-th category from type: answer Q_n: n-th category from type: question
T_n: n-th category from type: tell.

The ontology for maritime transport is structured into two levels: the navigation ontology and the communication ontology. It is important to distinguish between these two elements because they contain completely different issues from different fields. We also distinguish between functions that encode (f) and decode (g) messages in the ontology. The navigation ontology is based on a specified maritime vocabulary so systematized that it provides full unambiguous understanding of the message. The communication ontology takes into account the processes of obtaining and sharing information, as well as negotiations conducted between traffic participants that are actually carried out. It includes the type of message, e.g., question, request, response, and defines its structure, distinguishing the header and body of the message. They are needed, among other things, for the unambiguous interpretation of the sender and recipient of the message. The construction of the ontology was built using the Proteg'e program. It presents the structure of individual ontologies being created by defining the above hierarchy classes. The entire ontology can be represented as a tree structure (Figure 3.1), which naturally reflects the hierarchy of data (physical and abstract objects, concepts, etc.).

When assessing the effectiveness and correctness of the ontology in examples of ship encounters [8,9], the following criteria were used: the unambiguity of the communication, its effectiveness, and the time (of sending, receiving, and interpreting messages). A significant improvement in the clarity of communication was observed and, above all, the method's visible effectiveness was demonstrated, when potentially dangerous situations (in reality resulting in collisions) were solved without any problems. It was observed that the participants in the research experiment had to be thoroughly familiarized with the ontology interface (as a system not used on ships so far) to shorten the time of sending messages. Regarding the time required to receive and interpret messages, no impact on the communication time was found. All analyzed collision situations were solved by maintaining safe distances and correctly applying the COLREGs (Convention on the International Regulations for Preventing Collisions at Sea) regulations.

The coding and decoding of messages are based on functions that generate messages. These functions were created as the communication and navigational ontologies had to be combined. Based on the knowledge of information theory, set theory, and discrete mathematics, the necessary knowledge was gathered that allowed for formulating these functions. The function

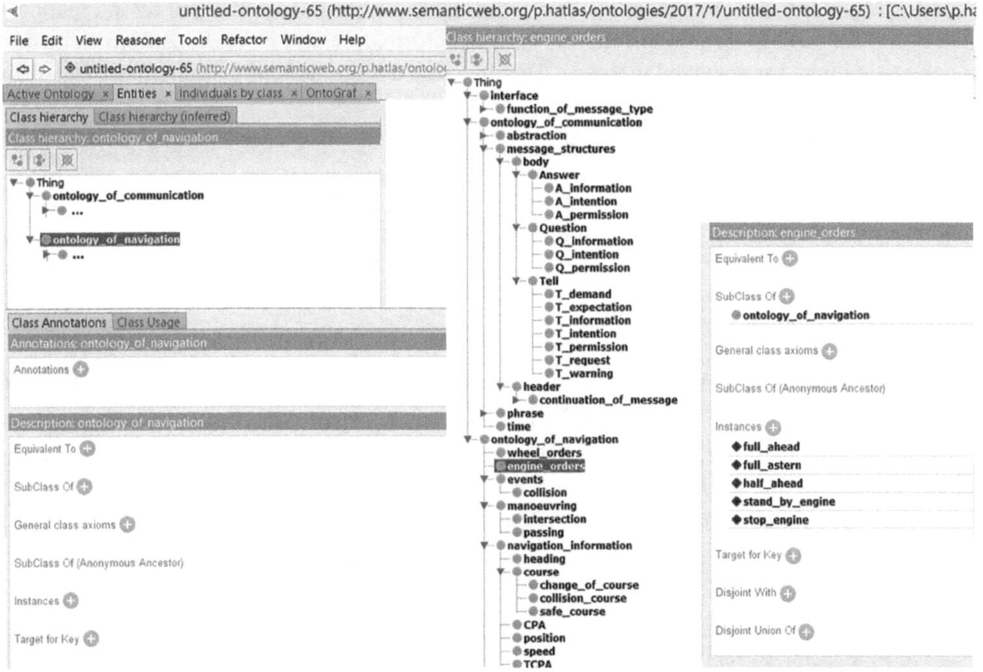

Figure 3.1 A mathematical tree presenting a fragment of the navigation and communication ontologies.

responsible for the generation of messages is called function f, while the function responsible for the unambiguous interpretation of messages is called function g.

The concept of information measure called entropy, used herein, is derived from the theory of information created by Shannon [10]. The theory uses the probability calculus for testing methods of transmission, gathering, and management of information.

The above theorem was used to determine the maximum length of generated messages. Based on knowledge from the concatenation of sequences and their properties, which are used to determine the method of knowledge representation, the authors create their own sequences of words (from a selected path) in the set X, where navigational terms are found. If $u = \{u_0, \ldots, u_{n-1}\}$ $v = \{v_0, \ldots, v_{m-1}\}$, these sequences are of length $\boldsymbol{n}$ and $\boldsymbol{m}$, respectively, their concatenation $u^\frown v$ is a sequence of length $\underline{\boldsymbol{n+m}}$, given by:

$$u^\frown v = \{u_0, \ldots, u_{n-1}, v_0, \ldots, v_{m-1}\} \tag{3.3}$$

Each word in message K (the message body) should be assigned to one of the defined ontologies: communication or navigation. The words assigned to the ontology of communication are collected in set Y. The words assigned to the ontology of navigation are collected in set X. If more than one element from set Y is chosen, they should be verified for possible contradictions. If they are contradictory, the function returns an error indicating that the message is incorrectly formulated (incomprehensible or ambiguous); otherwise, the function returns a pair of sets:

$$\left(y_n, X_k\right) = \left(y_n, x_{k1}, x_{k2}, \ldots, x_{kj}\right) \tag{3.4}$$

Function *f* connects one element of set *Y* with a selected element or elements of set *X*, creating *K*. Function g defines the meaning of the transmitted information and indicates actions to be done in connection with the received message. This function assigns a combination of elements from sets *Y* and *X* to the received message K_i.

The Cartesian product $Y \times X$ of sets *Y*, *X* is a set of all ordered pairs (y,x) such that $y \in Y$ and $x \in X$. Hence, $Y \times X = \{(y,x) : y \in Y \wedge x \in X\}$.

As the function *f* is determined on non-empty sets *Y*, *X* $(Y \neq \varnothing \wedge X \neq \varnothing)$ with values in a non-empty set *K* $(K \neq \varnothing)$, $f : Y \times X \rightarrow K$ meets the following conditions:

$$\Lambda_{y \in Y} \wedge \Lambda_{x_1, x_2 \in X} \left[(y, x_1) \in f \wedge (y, x_2) \in f \right] \Rightarrow y = y,$$

e.g.: {*T*_information, course}—state the course and {*T*_information, position}—state the position; the domain is: $Y \times X = \left\{ (y_n, X_k) : y_n \in Y \wedge X_k \subset X \right\}$, that is, a set of all elements belonging to set *X*, which are in relation with at least one element from set *Y*): the set of values of the function makes up the set *K*.

The functions *f* and *g* are expressed by Equations (3.5) and (3.6):

$$f : Y \times X \rightarrow K \quad f(y_n, X_k) = K_i \tag{3.5}$$

$$g : K \rightarrow Y \times X \quad g(K_i) = (y_n, X_k) \tag{3.6}$$

where
$Y = \{y_1, y_2, \ldots, y_l\}$—set of messages with the assigned category, $(1 \in N)$,
$y_n \in Y$—selected type with the message category assigned to it, $(1 \leq n \leq l)$
$X = \{x_1, x_2, \ldots, x_m\}$—set of navigational terms, $(m \in N)$,
$X_k = \{x_{k1}, x_{k2}, \ldots, x_{kj}\} \subset X$—set of terms in *k*-th message,
k—number of the message sent $(k \in N)$,
K—set of messages,
K_i—*i*-th message from set K,
$K_i = \{s_{in}, s_{i1}, s_{i2}, \ldots, s_{ij}\}$—individual words, used in the message, $(i, j, n \in N)$

where
s_{in}—*n*-th word in *i*-th message K_i from set *Y*
s_{ij}—*j*-th word in *i*-th message K_i from set *X*.

Example:

The Cartesian product of sets *Y*={*A*_intention, *T*_information} and *X*={course, position, passing} contains six ordered pairs:

Y×*X*={(*A*_intention, course), (*A*_intention, position), (*A*_intention, passing) (*T*_information, course), (*T*_information, position), (*T*_information, passing)}.

A Cartesian product of set *Y* cannot be formed with the same set *Y*, i.e., *Y*×*Y*, because the created message *K* contains only a single type and category. However, we can create a Cartesian product of set X with itself, i.e., for the same path of words, e.g., X= {course, alter course, to port, collision course}, where according to the definition given earlier the ordered pairs are $X \times X$= {(course, alter course,), (course, collision course), (alter course, to port)}.

Example of message 1 (K_1):

$$
\begin{aligned}
&f(y_1, X_1) = K_1 \\
&y_1 = \{T\_\text{information}\} \\
&X_1 = \{x_{11}, x_{12}, x_{13}\} \\
&X_1 = \{(\text{course}), (\text{altercourse}), (\text{to starboard})\}, \\
&\text{where } x_{11} = \text{course} \quad x_{12} = \text{alter course} \quad x_{13} = \text{to starboard}
\end{aligned} \tag{3.7}
$$

Example of message 2 (K_2):

$$
\begin{aligned}
&f(y_2, X_2) = K_2 \\
&y_2 = \{A\_\text{information}\} \\
&X_2 = \{(\text{engine orders}), (\text{stop engines})\}, \\
&\text{where } x_{21} = \text{engine orders} \quad x_{22} = \text{stop engines}
\end{aligned} \tag{3.8}
$$

The communication continues by using data coming from devices cooperating with navigational systems of the examined ships (ECDIS (electronic chart display and information system), ARPA (automatic radar plotting aids)). These devices are commonly referred to as sensors of navigational systems. Based on the information from the sensors, the communicating operator sends messages and unequivocally interprets the received messages.

3.4 Application of ontology in MASS

Currently, the exchange of information between two ships is done in a standardized manner, according to strict procedures and established message structures. The ontology of communication, which constitutes a different approach to the system of communication, is intended to help to unequivocally formulate and unambiguously interpret messages sent between ships. The authors' research so far has been based on semi-automatic communication, which involves the navigator's participation (Figure 3.2).

The research concept for solving the problem related to human errors is the application of artificial intelligence algorithms and machine learning solutions in the field of "Text mining" [11] also referred to as "natural language processing" [12] to automate communication between ships with an implemented autonomous navigation solution, as well as to automate communication between ships with a human navigator and a ship with an autonomous navigation system. The application of the communication models for the solution described in the article takes

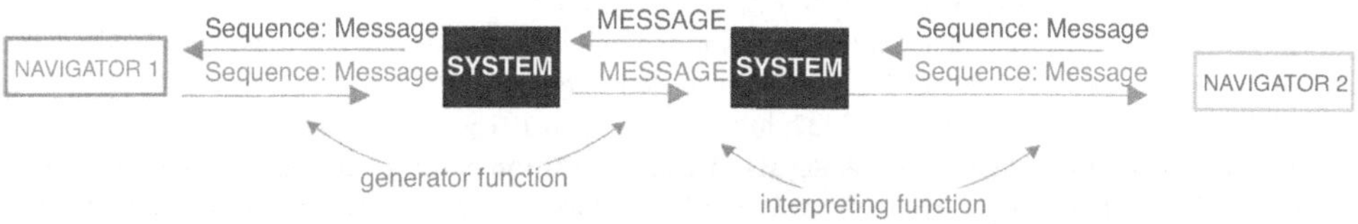

Figure 3.2 Automatic communication diagram: system – system including the navigator.

into account bidirectional communication between the human navigator and the autonomous system and between the autonomous system and the human. Autonomy in maritime transport allows for improved efficiency, safety, and sustainability, but the role of the human remains crucial for supervision and intervention. The report [13] emphasizes the importance of placing the human at the center of autonomy, advocating the integration of applied human intelligence into MASS technologies. This human-centered approach ensures that the evolution of autonomous systems in maritime transport will enhance, not replace, the roles of critical analysis and decision-making.

3.5 Conclusion

Currently, information exchange in the ship-to-ship relationship occurs in a standardized manner, according to strictly defined schemes and established message structures. The ontology proposed by the author, which is a different approach to the communication system, is to help in the unambiguous formulation and interpretation of messages sent between ships.

For verification, simulation studies involving expert navigators and survey studies were conducted. The ECDIS simulator was used in the simulation studies. This enabled the implementation of selected and previously prepared ship encounter scenarios. The study participants independently decided on the scope and course of communication. The studies confirmed the effectiveness of formulating statements in the analyzed situations in the communication process. Expert studies (surveys) involving expert navigators enabled the assessment of the actual communication process and the verification of the semi-automatic communication system. Most respondents considered the proposed communication method useful and expressed a positive opinion of its use.

It is possible that the use of the developed navigation and communication ontology for the automatic communication system in maritime navigation on autonomous ships will bring tangible results and contribute to increasing the level of safety.

References

1. Autonomous Ships Market Share, Size, Trends – [2022–2030], MarketsandMarkets. https://www.marketsandmarkets.com/Market-Reports/autonomous-ships-market-267183224.html
2. Pietrzykowski Z., Banaś P., Wołejsza P., Hatłas P. Subontologia komunikacji w automatyzacji procesów wymiany informacji i negocjacji na morzu. *Logistyka*, 6, 2014. https://yadda.icm.edu.pl/baztech/element/bwmeta1.element.baztech-a7d181b8-b6cb-4533-a696-9bbebe8c097a
3. Standard Marine Communication Phrases. Publishing Department of the Maritime University of Szczecin, Szczecin, 2009.
4. What is an Ontology? https://plato.stanford.edu/entries/ontology-is/
5. Ontology Learning for the Semantic Web. https://doi.org/10.1109/5254.920602
6. Scholz E. Herman Weyl on the concept of continuum. In V. F. Hendricks, S. A. Pedersen, K. F. Jørgensen (eds.), *Proof Theory: History and Philosophical Significance*. Dordrecht: Springer Netherlands, 2000, pp. 195–217. https://doi.org/10.1007/978-94-017-2796-9
7. Krajewski. https://wydawnictwa.ptm.org.pl/index.php/wiadomosci-matematyczne/article/view/3820
8. Hatłas-Sowińska P., Wielgosz M., Ontology based approach in solving collision situations at sea, *Ocean Engineering*, 260, 111941, 2022. https://doi.org/10.1016/j.oceaneng.2022.111941
9. Hatłas-Sowińska P., Wielgosz M. Analysis of communication ontology between sea-going vessels in real collision situations. *Zeszyty Naukowe Akademii Morskiej w Szczecinie*, 69(141), 2022. https://doi.org/10.17402/502

10. Shannon C., Weaver W. *The Mathematical Theory of Communication.* Nokia Bell Labs, 1948.
11. The Text Mining Handbook: Advanced Approaches in Analyzing Unstructured Data. https://www.researchgate.net/publication/200504395
12. *Handbook of Natural Language Processing.* https://doi.org/10.1162/coli.2000.27.4.602
13. Maritime Autonomous Surface Ships (MASS) | LR. https://www.lr.org/en/knowledge/research-reports/2024/maritime-autonomous-surface-ships/

Chapter 4

Autonomous harbor tug formation control system

Wojciech Koznowski and Andrzej Łebkowski

4.1 Introduction

Recent trends in the field of unmanned vessels are concentrating on automation of navigation, a task that was until recently reserved for the ship's crew. Achievements in the field of robotics and artificial intelligence systems have made it possible to create vessels capable of autonomously performing complex operations. A fundamental element of autonomous navigation is the optimization of the process of determining the vessel's route. A specific group of vessels are port tugs, whose function is to support larger vessels during port operations. The characteristics of their work, concentrated in a specific port area [1], generate special requirements in the context of route planning. The port space is characterized by high traffic intensity of various types of vessels, which, combined with limited operational space, creates an increased risk of collisions. Tugs must operate effectively in this environment, combining maneuvering precision with safe navigation. The port infrastructure, including a system of channels, quays, and breakwaters, is an additional limitation of the maneuvering space that must be taken into account in the route planning process.

There is a growing trend to implement electric and hybrid drives in port tugs [2]. This technological transformation, despite significant environmental benefits, introduces new operational challenges. The limited capacity of energy storage systems compared to conventional combustion units requires particularly precise planning of operations. Effective energy management becomes a key parameter in the route optimization process, where inefficient maneuvers can lead to premature depletion of power sources. The issue of route planning for autonomous seagoing units requires a comprehensive approach [3], taking into account many criteria, such as energy economy, time efficiency, meteorological factors, and priority aspects of the unit's safety and the marine ecosystem.

Modern computational methods allow for various approaches to route optimization. Possible methods include graph algorithms, potential and vector methods, evolutionary techniques, neural networks, and hybrid solutions. Graph methods are based on discretization of the navigation space and assigning potential positions of the unit for which the route is being determined to the graph vertices and admissible transitions between these positions to the graph edges. An example of a graph method is the method using the Voronoi diagram [4], where the navigation area is divided into regions based on the distance from obstacles. The method using the rapidly exploring random tree (RRT) algorithm uses probabilistic exploration techniques [5–8].

The approach using artificial potential fields (APF) operates on the concept of force fields, both attractive and repulsive [9–14]. Popular solutions are evolutionary methods, including ant colony optimization (ACO) [10, 15, 16] and methods based on Genetic Algorithms (GA).

DOI: 10.1201/9781003686088-4

References [17–21] implement optimization mechanisms inspired by the biology of living organisms.

Artificial neural networks [22–25] and modified classical graph algorithms [4, 26–29] are other important optimization tools. The development of computing power has also enabled the implementation of reinforcement learning methods [15, 30–34].

This paper presents an innovative methodology for energy optimization of autonomous tugboat routes, integrating an evolutionary algorithm with radar fitness assessment. The proposed solution focuses on minimizing energy consumption while maintaining high standards of navigational safety.

4.2 Calculating a passage route with evolutionary algorithm

Adaptive mechanisms occurring in nature are the basis for the functioning of evolutionary algorithms, where the key element is the selection pressure favoring the best-adapted individuals. Analogously to biological processes, artificial evolution uses selection mechanisms to optimize solutions. In the context of GA, which are a widespread class of evolutionary algorithms, the fundamental structure is the genome of an individual. In contrast to biological genomes, the structure of genetic information in evolutionary algorithms is reduced to elements that are essential from the perspective of the problem being solved. The dominant paradigm of genome representation is binary coding, where genetic information is recorded in the form of chromosomes containing genes with binary values. Implementing a genetic algorithm in the context of vessel navigation requires the appropriate transformation of navigation parameters into binary form. Individual control variables, such as the course of the vessel, speed, or coordinates of turning points, are represented by bit sequences with a length determined by the required precision of the representation. For example, a digital representation of a course with a resolution of 2048 values requires the use of 11 bits of information.

The optimization process in a genetic algorithm is carried out through a sequence of operations: population initialization, fitness evaluation, selection, genetic recombination, and mutation. Initialization involves creating an initial population consisting of randomly generated individuals, where the population size typically ranges from several dozen to several hundred individuals. Each genome represents a potential solution to the navigation problem.

The quality of individual solutions is assessed using a fitness function, which quantifies the degree to which the assumed optimization criteria are met. The selection process most often uses the roulette method, where the probability of selecting an individual is correlated with the value of its fitness function. This mechanism favors higher quality solutions, increasing their share in the genetic pool of the next generation. Genetic recombination is carried out through a crossover operation, where the genetic material of selected individuals is combined to create new solutions.

The basic implementation of this operation involves a single-point intersection of parental chromosomes and the exchange of the resulting fragments. A complementary mechanism is mutation, which involves stochastic modification of individual genes with a specified probability (typically a few percent). This operation ensures diversification of the gene pool and prevents convergence to local extremes. The evolutionary process is implemented iteratively, where each generation includes a complete cycle of genetic operations. The algorithm continues to work until the defined criteria for completing the calculations are met, such as achieving a satisfactory level of fitness or performing a specified number of iterations. In the context of navigation route optimization, the algorithm operates on the space of solutions defined by the initial position of the unit, the destination point, and a set of navigation constraints (static and dynamic).

The initial population of routes is generated as a set of broken lines connecting the starting point with the destination, where each segment is assigned a specific speed. The parameterization of the algorithm includes the population size, the ranges of the turning point coordinates, and the permissible speed ranges on individual route sections.

Effective implementation of the genetic algorithm in the context of navigation route optimization requires comprehensive integration of navigation constraints both at the stage of generation and evaluation of individuals. The generated solutions, despite their initial stochastic nature, are subject to rigorous verification for compliance with applicable navigation regulations, including collision avoidance and adherence to designated waterways. The verification process of each individual includes an analysis of compliance with static and dynamic constraints. Disqualification or significant penalization in the assessment of fitness occurs when any route segment collides with terrain obstacles, penetrates zones closed to navigation, or generates a risk of collision with other units operating in the water. The catalog of static restrictions includes the following elements:

- Terrain formations and associated shallows
- Permanent elements of navigation infrastructure (buoys, breakwaters, and mooring piles)
- Objects posing a navigational hazard (wrecks and underwater obstacles)
- Defined navigation zones with specific temporary or permanent restrictions and fairways with predefined movement parameters

The category of dynamic restrictions includes all mobile objects, primarily floating units (referred to as foreign ships), as well as other moving elements such as ice formations or moving cargo (e.g., containers). Static restrictions, characterized by the invariability of position and dimensions over time, can be digitized in the form of a digital map, constituting a database for the algorithm in the process of verifying the admissibility of planned routes. To increase the safety margin, restriction areas can be extended with an additional buffer zone [16]. Dynamic constraints are characterized by a certain level of uncertainty resulting from the characteristics of object detection and tracking systems, such as:

- radar with ARPA module
- LIDAR systems [11,23]
- optical sensors operating in the range of visible and infrared light [11]
- Automatic Identification System (AIS) [35]

The positioning of the own ship is carried out using GNSS systems, including GPS [11], where both position and speed data (from GNSS systems or on-board logs) are subject to a certain measurement error. In the case of identifying an individual that does not meet the established criteria, various correction strategies are possible, such as:

- Implementation of a penalization system in the goal function, reducing the value of adaptation of routes violating constraints,
- Complete elimination of unacceptable routes and replacing them with new individuals generated randomly or through genetic operations on the existing population.

The evaluation of an individual's fitness in the context of navigation route planning should take into account two fundamental aspects: The economic efficiency of the route is determined by the ratio of the length of the designated route to the theoretical distance between points, the total passage time, and the characteristics of the required maneuvers (number and their intensity).

The requirement of the route determination process is to determine the location of the starting and destination points. The total length of the route L_{tot}, consisting of k segments, is the sum of the lengths of the individual component segments.

$$L_{\text{tot}} = \sum_{i=1}^{k} L_i \tag{4.1}$$

where:
L_i – length of the i-th segment,
k – total number of segments.

Route implementation requires a sequence of course changes at defined turn points, where the total value of all direction changes, expressed in angular units (degrees), is defined by the parameter θ_{tot}.

$$\theta_{\text{tot}} = \sum_{n}^{k-1} \theta_n \tag{4.2}$$

where:
θ_n – course change after completing the n-th route segment,
k – number of route segments.

Assuming a constant speed within each segment of the route, the total transit time (t_{tot}) can be expressed as the sum of the times of completing individual sections. The time to complete each segment is the product of its length and the speed assumed for it.

$$t_{\text{tot}} = \sum_{i}^{k} L_i \cdot v_i \tag{4.3}$$

where:
L_i – length of the i-th segment,
v_i – speed of completing the i-th segment.

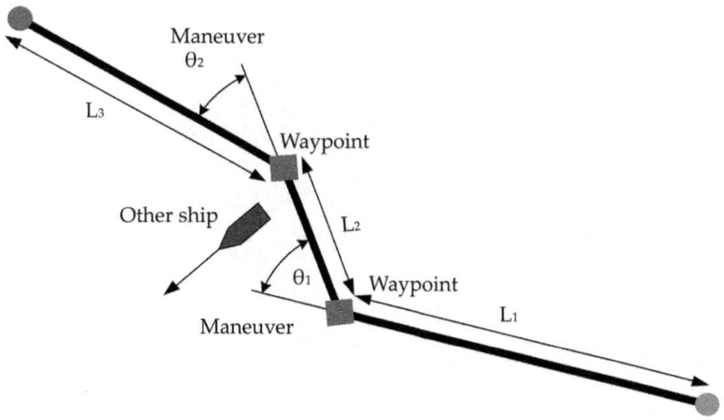

Figure 4.1 Diagram of an example route with important parameters marked.

The diagram shown in Figure 4.1 illustrates an example route connecting two points, consisting of three segments separated by two waypoints. The diagram also shows the presence of a foreign vessel, whose trajectory intersects the planned route. The need for appropriate interaction with this object results from international regulations specified in the COLREG Convention (Convention on the International Regulations for Preventing Collisions at Sea) [36], defining protocols for the behavior of vessels during mutual encounters in the water.

4.3 Determining the local transition route using an evolutionary algorithm with a radar-based fitness assessment method

The proposed methodology is an evolution of the standard genetic algorithm by implementing an innovative system for determining fitness based on the analysis of radar charts. The concept of this method is based on the evaluation of the radar chart surface generated by parameters characteristic of a given navigation route, represented by an individual in an evolving population. A significant advantage of the presented approach is the flexibility in introducing many evaluation criteria, which allows the algorithm to be adapted to the specific requirements of local route optimization. Figure 4.2 presents a visualization of three alternative navigation routes, while Figure 4.3 illustrates the corresponding radar charts, representing the degree of fitness of individual routes.

The presented five-parameter radar chart is the basic configuration of the evaluation system. The system is characterized by the potential of expansion with additional evaluation criteria, such as variations of the unit's speed or the influence of environmental factors (sea currents, weather conditions, and bathymetric characteristics of the water body) on the characteristics of the unit's swimming resistance.

The analysis of the route marked in red indicates its minimum total length, but it is characterized by a significant number of turning points, implying intensive course changes. Despite

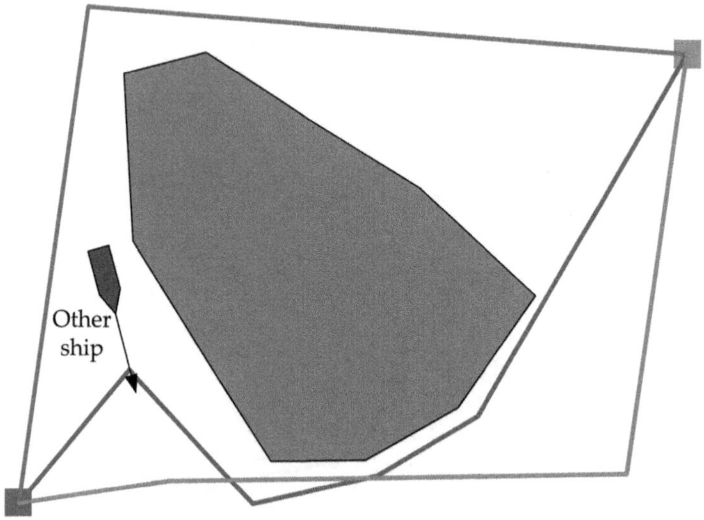

Figure 4.2 View of three sample navigation routes to be evaluated.

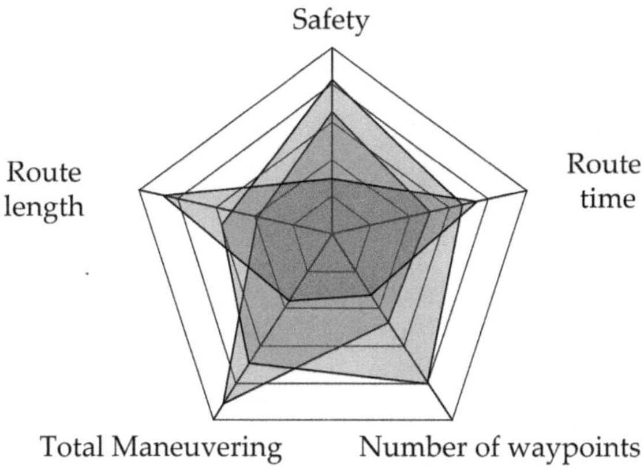

Figure 4.3 Radar plots of three sample navigation routes to be evaluated.

the shortest passage time, this route presents a reduced level of safety resulting from the close proximity to navigational obstacles, both static (land) and dynamic (other ships).

The route marked in blue is characterized by a reduced number of turning points, but the maneuvers performed are characterized by significant course change angles. This results in an extension of both the passage route and the total time required to complete the route.

The route marked in green is distinguished by a single turning point and, despite a moderate extension compared to the blue variant, eliminates the need to perform an anti-collision maneuver by conducting an alternative route in relation to a stationary obstacle in the form of an island.

Using the radar assessment of adaptation, the green route achieves the highest value of the objective function, as it has the largest surface area on the radar graph. The blue route presents an average level of adaptation, while the red route is characterized by the lowest value of the objective function. The applied visualization methodology enables effective comparative analysis of results for different sets of evaluation rules, facilitating the process of selecting the optimal parameter configuration in the context of specific navigation scenarios.

4.4 Research

The effectiveness of the proposed algorithm was verified based on four navigation scenarios, implemented in restricted water conditions, with static and dynamic obstacles. In all the figures presented in this chapter, north was assumed to be in the direction of the top of the figure.

A summary of the parameters related to the distance covered and course changes is included in Tables 4.1 and 4.2, placed at the end of the chapter. In the first test scenario, illustrated in Figure 4.4, the navigation space is characterized by restrictions from the north-west and east, without obstacles in the center.

The analyzed tugboat, marked with a green symbol, is located in the southern part of the water body. The dashed line connecting the initial position with the target point shows the initial course vector. There are five foreign units in the operational space, marked with red symbols, with marked course vectors and target points. An important assumption is the lack of maneuverability of foreign units, which function only as moving navigational obstacles. The graphical representation of the solution to the first scenario is shown in Figure 4.5.

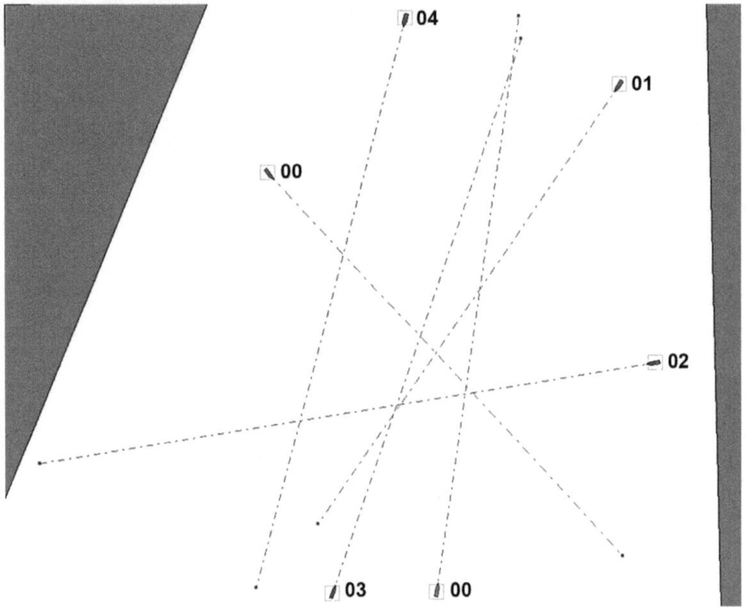

Figure 4.4 Scenario 1, constricted waters with five other ships.

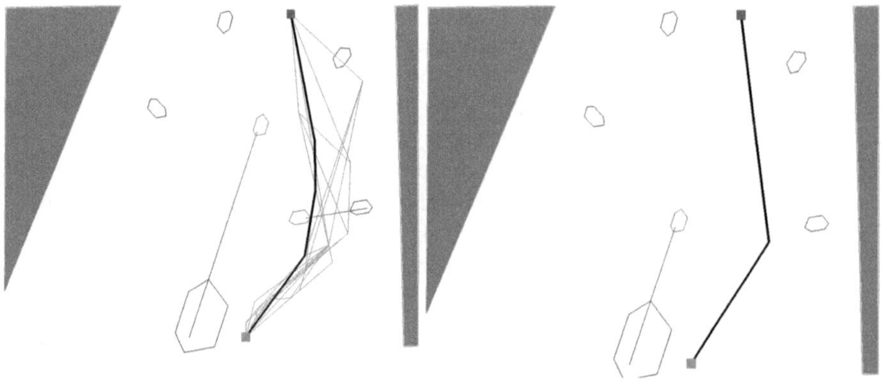

Figure 4.5 Solutions of scenario 1: (a) with classic fitness calculation, (b) with improved radar plot fitness calculation.

The comparative analysis of the routes generated by both versions of the evolutionary algorithm indicates similarity; however, the route (a) obtained by the classical fitness assessment method is characterized by two additional turning points. The algorithm using the radar method generated a shorter route, with a single turning point, optimized for safe passing the foreign vessel by its stern, which is the preferred anti-collision maneuver compared to passing in front of the bow. The proposed route requires a reduction of course changes by 18.7% compared to the classical solution. Figure 4.5 introduces a representation of the security domains in the form of hexagons, where the blue color indicates the initial positions of the foreign units, and the red color illustrates the position at the points of potential approach or intersection of the routes.

The second scenario, presented in Figure 4.6, is characterized by increased complexity of the operational space by introducing two islands in the central part of the restricted water area and modifying the courses of five foreign units.

The tugboat begins the operation in a similar orientation to the first scenario, but its planned route intersects the track of a larger vessel moving from the north, and it is necessary to bypass the eastern island (leftmost, green triangular feature in Figure 4.7).

The algorithm using the classical fitness assessment method proposed a route containing two turning points, as opposed to the single point in the radar algorithm solution. The classical

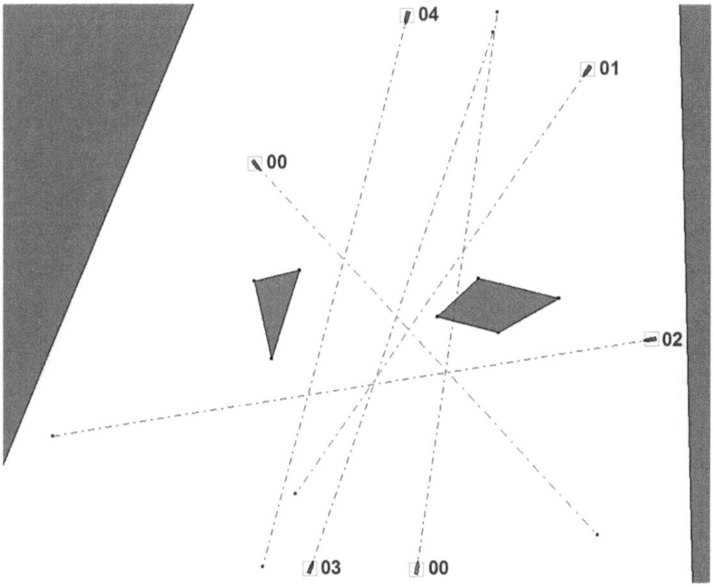

Figure 4.6 Scenario 2, constricted waters with 5 other ships and 2 islands.

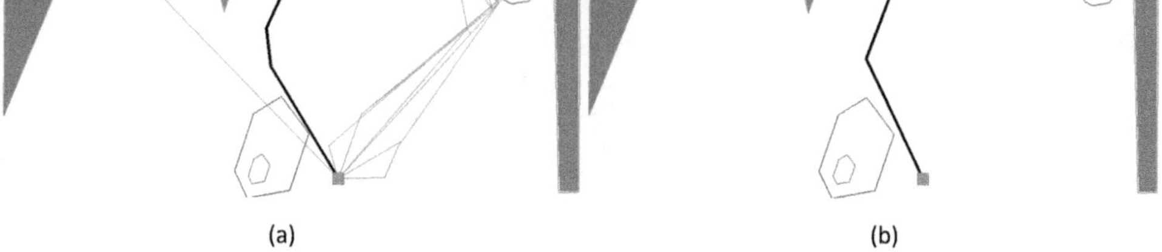

Figure 4.7 Solutions of Scenario 2: (a) with classic fitness calculation, (b) with improved radar plot fitness calculation.

solution is characterized by a longer total route length and higher maneuvering intensity. The first turning point is located at a significant distance from the foreign unit heading southwest ensuring safe separation at the intersection of the routes. The route generated by the radar assessment algorithm, despite having a shorter approach distance to this unit, still maintains a safety margin multiple times the length of its safety domain.

The third scenario, illustrated in Figure 4.8, presents a complex navigation space containing 8 islands.

In this case, the radar algorithm again proposed a single-turnpoint route, compared to two turnpoints on the route calculated by the algorithm with classical fitness assessment. The route proposed by the improved algorithm was also shorter and required a smaller total course change (Figure 4.9).

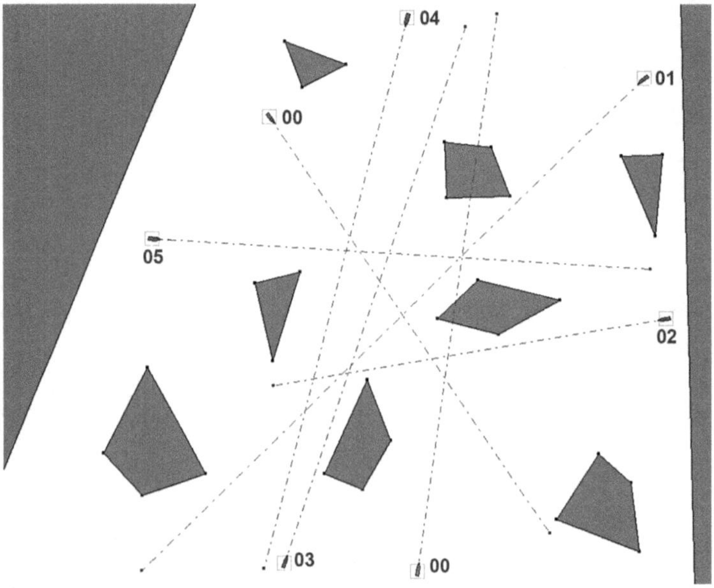

Figure 4.8 Scenario 1, constricted waters with 5 other ships and 8 islands.

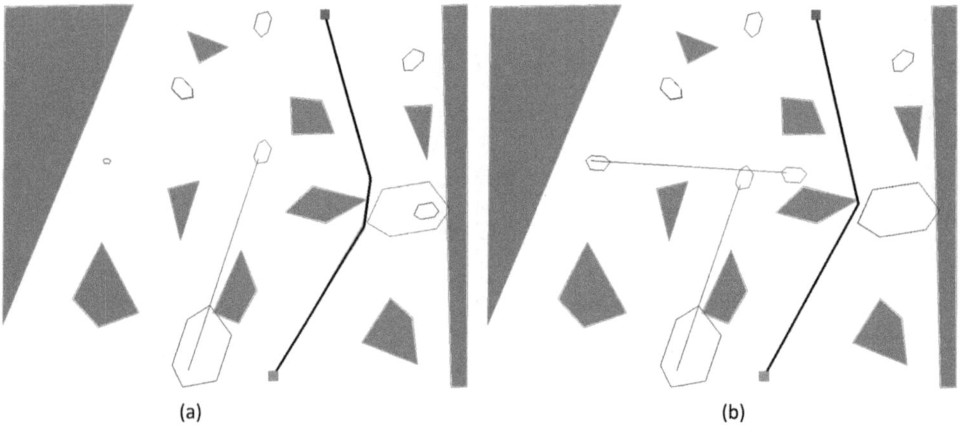

Figure 4.9 Solutions of Scenario 3: (a) with classic fitness calculation, (b) with improved radar plot fitness calculation.

The fourth test scenario was carried out in a newly defined operational space, limited by landmasses to the east and west, with the port located in the central part. The navigation space was additionally organized by introducing fairways around the port, separating traffic heading to the south to move on the fairway located to the west (on the left side of Figure 4.10) and the traffic heading to the north to move on the fairway one the east (on the right side of Figure 4.10). In the operational space, in addition to the towing unit, six foreign units were placed, one of which carries out the operation at low speed inside the port basin (Figure 4.10).

The algorithm using the classic method of assessing suitability generated a route characterized by a close distance to the port quays and two additional turning points compared to the radar algorithm solution. In this case, the route determined by the radar algorithm, despite a slightly larger total length, is characterized by a significantly reduced total course change and a significantly larger safety margin with respect to obstacles, which translates into a higher assessment of navigational safety (Figure 4.11).

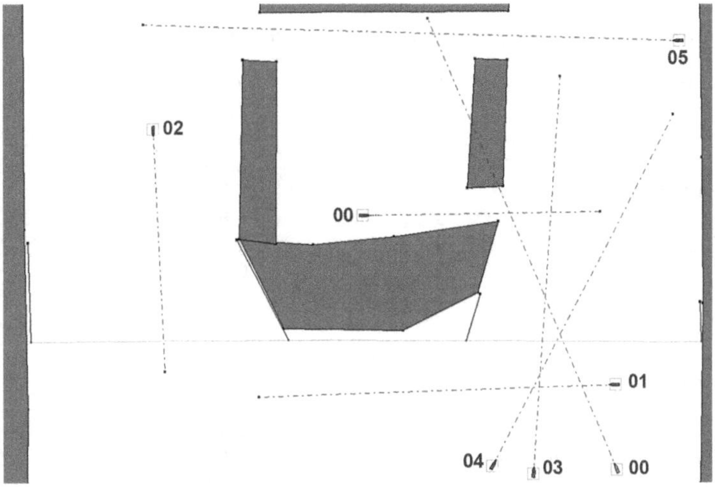

Figure 4.10 Scenario 1, constricted waters with 6 other ships and a port.

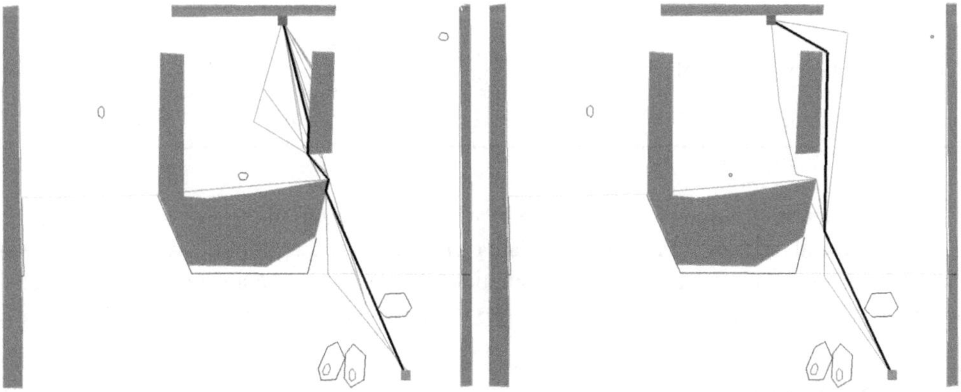

(a) (b)

Figure 4.11 Solutions of scenario 4: (a) with classic fitness calculation, (b) with improved radar plot fitness calculation.

Table 4.1 Summary of every scenario with route length and course change data

Scenario, algorithm type	*Route length (nm)*	*Route difference (%)*	*Course change (deg0)*	*Course change difference (%)*
Scenario 1, Classic fitness method	7.90	–	47.61	–
Scenario 1, Radar fitness method	7.85	−0.67	40.11	18.68
Scenario 2, Classic fitness method	8.14	–	56.95	–
Scenario 2, Radar fitness method	7.90	−3.05	46.56	22.32
Scenario 3, Classic fitness method	7.99	–	47.25	–
Scenario 3, Radar fitness method	7.92	−0.87	44.77	5.52
Scenario 4, Classic fitness method	7.74	–	146.09	–
Scenario 4, Radar fitness method	8.17	+5.26	87.93	66.15

Table 4.2 Route segment data for presented scenarios

Scenario, algorithm type	*Waypoint number*	*Section length (nm)*	*Course (deg)*
Scenario 1, Classic Fitness Method	1	2.32	36.7
	2	1.47	9.8
	3	1.18	359.0
	4	2.93	349.1
Scenario 1, Radar Fitness Method	1	3.04	33.1
	2	4.81	352.9
Scenario 2, Classic Fitness Method	1	2.16	328.2
	2	0.64	354.6
	3	5.34	25.2
Scenario 2, Radar Fitness Method	1	2.22	334.4
	2	5.68	20.9
Scenario 3, Classic Fitness Method	1	3.57	31.6
	2	0.97	7.7
	3	3.45	344.4
Scenario 3, Radar Fitness Method	1	4.01	29.1
	2	3.91	347.0
Scenario 4, Classic Fitness Method	1	4.05	336.4
	2	0.25	11.3
	3	0.67	320.4
	4	0.54	3.2
	5	2.22	345.7
Scenario 4, Radar Fitness Method	1	3.23	334.1
	2	3.58	1.1
	3	1.35	300.2

Table 4.1 presents a comprehensive summary of the total lengths of the routes generated by both versions of the algorithm, along with the relative change in the length of the route determined by the radar method in relation to the classical solution. The navigation parameterization of the generated routes, including course values and distances for individual segments, is presented in Table 4.2.

4.5 Summary

The conducted research has shown the high efficiency of the evolutionary algorithm using the radar method for assessing the adaptation in the process of optimizing navigation routes for autonomous towing units operating in restricted waters and port spaces.

Comparison with the classic evolutionary algorithm revealed the following advantages of the proposed solution:

- The algorithm with radar assessment of adaptation is characterized by a reduction in the number of turning points by 1–2 compared to the classic evolutionary algorithm in each analyzed scenario.
- Significant minimization of the total change of course, especially visible in the fourth scenario (66.15%), and its noticeable reduction in the remaining scenarios, translates directly into increased energy efficiency.
- A characteristic feature of the generated routes is an increased safety margin with respect to obstacles, which is particularly important in the context of the fourth scenario.
- The total length of the generated routes remains at a comparable or more favorable level, with variations ranging from −3.05% to +5.26%.

Particularly noteworthy is the high efficiency of the algorithm in scenarios of increased complexity (scenarios 3 and 4), where, despite the increased number of navigation constraints, the system effectively generated routes characterized by a higher level of safety while maintaining a comparable total length.

The proposed methodology for radar fitness assessment in the context of evolutionary algorithms presents itself as a prospective solution in the domain of route planning for autonomous towing units, offering the following key benefits: minimization of the number of required maneuvers, ensuring an increased standard of operational safety and optimization of the route length.

- The proposed directions for further research include:
- Implementation and validation of the algorithm in operational conditions
- Optimization of the radar method parameterization
- Extension of the optimization criteria to include detailed aspects of energy efficiency

Integration with traffic prediction systems [37] and communication platforms proposed within the e-navigation strategy [38] is being developed under the auspices of the International Maritime Organization.

References

1. Przybylowski, A.; Suchanek, M.; Miszewski, P. COVID-19 pandemic –impact on a global liner shipping company employee work digitalization. *TransNav-International Journal on Marine Navigation and Safety of Sea Transportation* **2022**, *16*, 759 765. https://doi.org/10.12716/1001.16.04.18
2. Muc, A. Single-phase cascade inverter controlled by signals calculated on the basis of the Haar wavelet. *Electrotechnical Review* **2023**, *1*, 232–235. https://doi.org/10.15199/48.2023.09.45
3. CZECH, P. Autonomous vehicles: basic issues. *SJSUT.ST* **2018**, *100*, 15–22. https://doi.org/10.20858/sjsutst.2018.100.2
4. Niu, H.; Lu, Y.; Savvaris, A.; Tsourdos, A. Efficient path planning algorithms for unmanned surface vehicle. *IFAC-PapersOnLine* **2016**, *49*, 121–126. https://doi.org/10.1016/j.ifacol.2016.10.331
5. Cao, S.; Fan, P.; Yan, T.; Xie, C.; Deng, J.; Xu, F.; Shu, Y. Inland waterway ship path planning based on improved RRT algorithm. *Journal of Marine Science and Engineering* **2022**, *10*, 1460. https://doi.org/10.3390/jmse10101460
6. Yu, J.; Chen, Z.; Zhao, Z.; Yao, P.; Xu, J. A traversal multi-target path planning method for multi-unmanned surface vessels in space-varying ocean current. *Ocean Engineering* **2023**, *278*, 114423. https://doi.org/10.1016/j.oceaneng.2023.114423

7. Zaccone, R.; Martelli, M.; Figari, M. A COLREG-compliant ship collision avoidance algorithm, 2530–2535. https://doi.org/10.23919/ECC.2019.8796207
8. Zhang, Z.; Wu, D.; Gu, J.; Li, F. A path-planning strategy for unmanned surface vehicles based on an adaptive hybrid dynamic stepsize and target attractive force-RRT algorithm. *Journal of Marine Science and Engineering* **2019**, *7*, 132. https://doi.org/10.3390/jmse7050132
9. Iswanto, I.; Ma'arif, A.; Wahyunggoro, O.; Imam, A. Artificial Potential Field Algorithm Implementation for Quadrotor Path Planning. *International Journal of Advanced Computer Science and applications* **2019**, *10*. https://doi.org/10.14569/ijacsa.2019.0100876
10. Lyu, H.; Yin, Y. COLREGS-constrained real-time path planning for autonomous ships using modified artificial potential fields. *Journal of Navigation* **2019**, *72*, 588–608. https://doi.org/10.1017/S0373463318000796
11. Mousazadeh, H.; Jafarbiglu, H.; Abdolmaleki, H.; Omrani, E.; Monhaseri, F.; Abdollahzadeh, M.-r.; Mohammadi-Aghdam, A.; Kiapei, A.; Salmani-Zakaria, Y.; Makhsoos, A. Developing a navigation, guidance and obstacle avoidance algorithm for an Unmanned Surface Vehicle (USV) by algorithms fusion. *Ocean Engineering* **2018**, *159*, 56 65. https://doi.org/10.1016/j.oceaneng.2018.04.018
12. Muc, A.; Kasprowicz, A.; Bielecka, A.; Mysiak, P.; Iwaszkiewicz, J. The influence of the reference potential on the operation of the pulse formation system of a 12-pulse controlled rectifier. *Electrotechnical Review* **2024**, *1*, 234–237. https://doi.org/10.15199/48.2024.10.48
13. Park, M.G.; Lee, M.C. A new technique to escape local minimum in artificial potential field based path planning. *KSME International Journal* **2003**, *17*, 1876–1885. https://doi.org/10.1007/BF02982426
14. Wang, H.; Ban, X. Research on autonomous collision avoidance method of unmanned surface vessel in the circumstance of moving obstacles. In *2018 37th Chinese Control Conference (CCC)*. 2018 37th Chinese Control Conference (CCC), Wuhan, 25–27 Jul. 2018; IEEE, 2018, pp. 501–506.
15. Li, L.; Wu, D.; Huang, Y.; Yuan, Z.-M. A path planning strategy unified with a COLREGS collision avoidance function based on deep reinforcement learning and artificial potential field. *Applied Ocean Research* **2021**, *113*, 102759. https://doi.org/10.1016/j.apor.2021.102759
16. Liu, Y.; Bucknall, R.; Zhang, X. The fast marching method based intelligent navigation of an unmanned surface vehicle. *Ocean Engineering* **2017**, *142*, 363–376. https://doi.org/10.1016/j.oceaneng.2017.07.021
17. Arzamendia, M.; Gregor, D.; Reina, D.G.; Toral, S.L. An evolutionary approach to constrained path planning of an autonomous surface vehicle for maximizing the covered area of Ypacarai Lake. *Soft Computing* **2019**, *23*, 1723–1734. https://doi.org/10.1007/s00500-017-2895-x
18. Fiskin, R.; Atik, O.; Kisi, H.; Nasibov, E.; Johansen, T.A. Fuzzy domain and meta heuristic algorithm-based collision avoidance control for ships: experimental validation in virtual and real environment. *Ocean Engineering* **2021**, *220*, 108502. https://doi.org/10.1016/j.oceaneng.2020.108502
19. Kim, H.; Kim, S.-H.; Jeon, M.; Kim, J.; Song, S.; Paik, K.-J. A study on path optimization method of an unmanned surface vehicle under environmental loads using genetic algorithm. *Ocean Engineering* **2017**, *142*, 616–624. https://doi.org/10.1016/j.oceaneng.2017.07.040
20. Seghir, M.M. Safe ship's control in a fuzzy environment using a genetic algorithm. *Solid State Phenomena* **2011**, *180*, 70–75. https://doi.org/10.4028/www.scientific.net/SSP.180.70
21. Xin, J.; Zhong, J.; Yang, F.; Cui, Y.; Sheng, J. An improved genetic algorithm for path planning of unmanned surface vehicle. *Sensors* **2019**, *19*. https://doi.org/10.3390/s19112640
22. Praczyk, T. Neural anti-collision system for autonomous surface vehicle. *Neurocomputing* **2015**, *149*, 559–572. https://doi.org/10.1016/j.neucom.2014.08.018
23. Sorial, M.; Mouawad, I.; Simetti, E.; Odone, F.; Casalino, G. Towards a real time obstacle detection system for unmanned surface vehicles, 1–8. https://doi.org/10.23919/OCEANS40490.2019.8962685
24. Wang, R.; Li, D.; Miao, K. Optimized radial basis function neural network based intelligent control algorithm of unmanned surface vehicles. *Journal of Marine Science and Engineering* **2020**, *8*, 210. https://doi.org/10.3390/jmse8030210

25. Wang, R.; Miao, K.; Li, Q.; Sun, J.; Deng, H. The path planning of collision avoidance for an unmanned ship navigating in waterways based on an artificial neural network. *Nonlinear Engineering* **2022**, *11*, 680–692. https://doi.org/10.1515/nleng-2022-0260
26. Casalino, G.; Turetta, A.; Simetti, E. A three-layered architecture for real time path planning and obstacle avoidance for surveillance USVs operating in harbour fields, 1–8. https://doi.org/10.1109/OCEANSE.2009.5278104
27. Naeem, W.; Irwin, G.W.; Yang, A. COLREGs-based collision avoidance strategies for unmanned surface vehicles. *Mechatronics* **2012**, *22*, 669–678. https://doi.org/10.1016/j.mechatronics.2011.09.012
28. Neumann, T. Vessels route planning problem with uncertain data. *TransNav-International Journal on Marine Navigation and Safety of Sea Transportation* **2016**, *10*, 459–464. https://doi.org/10.12716/1001.10.03.11
29. Singh, Y.; Sharma, S.; Sutton, R.; Hatton, D.; Khan, A. A constrained A* approach towards optimal path planning for an unmanned surface vehicle in a maritime environment containing dynamic obstacles and ocean currents. *Ocean Engineering* **2018**, *169*, 187–201. https://doi.org/10.1016/j.oceaneng.2018.09.016
30. Ai, B.; Jia, M.; Xu, H.; Xu, J.; Wen, Z.; Li, B.; Zhang, D. Coverage path planning for maritime search and rescue using reinforcement learning. *Ocean Engineering* **2021**, *241*, 110098. https://doi.org/10.1016/j.oceaneng.2021.110098
31. Chen, C.; Chen, X.-Q.; Ma, F.; Zeng, X.-J.; Wang, J. A knowledge-free path planning approach for smart ships based on reinforcement learning. *Ocean Engineering* **2019**, *189*, 106299. https://doi.org/10.1016/j.oceaneng.2019.106299
32. Guo, S.; Zhang, X.; Zheng, Y.; Du, A.Y. An autonomous path planning model for unmanned ships based on deep reinforcement learning. *Sensors* **2020**, *20*. https://doi.org/10.3390/s20020426
33. Perera, L.P. Deep learning toward autonomous ship navigation and possible COLREGs failures. *Journal of Offshore Mechanics and Arctic Engineering* **2020**, *142*, 2256. https://doi.org/10.1115/1.4045372
34. Zhao, L.; Szakas, T.; Churley, M.; Lucas, D. Multi-class multi-residue analysis of pesticides in edible oils by gas chromatography-tandem mass spectrometry using liquid extraction and enhanced matrix removal lipid cartridge cleanup. *J. Chromatogr. A* **2019**, *1584*, 1–12. https://doi.org/10.1016/j.chroma.2018.11.022
35. Gao, M.; Shi, G.-Y. Ship-collision avoidance decision-making learning of unmanned surface vehicles with automatic identification system data based on encoder – decoder automatic-response neural networks. *Journal of Marine Science and Engineering* **2020**, *8*, 754. https://doi.org/10.3390/jmse8100754
36. International Maritime Organization. *Convention on the International Regulations for Preventing Collisions at Sea*, London, 1972.
37. Kolakowski, P.; Rutkowski, G. The analysis of offshore industry transition – acceleration from oil and gas to wind. *TransNav* **2022**, *16*, 375–384. https://doi.org/10.12716/1001.16.02.20
38. International Maritime Organization. *E-navigation*. https://www.imo.org/en/OurWork/Safety/Pages/eNavigation.aspx

Chapter 5

The design process of collision avoidance algorithms for ships to be applied in the autonomous navigation system

Agnieszka Lazarowska

5.1 Introduction

The development of technology allows for the design and construction of autonomous vehicles for use in air, sea, road, and rail transport. An autonomous vehicle is a machine that uses different sensors and modern control methods to move and perform various tasks without human control. A recent example of such a solution is the Mayflower Autonomous Ship, also known as Mayflower 400 (MAS400). This autonomous research vessel was designed to cross the Atlantic Ocean without human assistance [1]. Another example of an innovative unmanned vehicle is an all-electric Vertical Take-off and Landing (eVTOL) multicopter aircraft produced by China's Guangzhou EHang Intelligent Technology [2].

The main system of unmanned and fully autonomous vehicles is the autonomous navigation system (ANS), which allows the craft to navigate in an environment without any human intervention. This system handles tasks such as environment perception, path planning and motion control. The main part of this system is responsible for collision avoidance (COLAV) with both static and dynamic obstacles.

The aim of this paper is to comprehensively present the process of designing and developing a COLAV algorithm for application in the ANS of Maritime Autonomous Surface Ships (MASS) or Unmanned Surface Vehicles (USVs). The rest of the paper is organized as follows: Section 2 presents the consecutive stages of the COLAV algorithm design process. The vital assumptions and constraints of the process are also briefly described there. Section 3 shows examples of solutions obtained with the use of COLAV algorithms for MASS and USVs. Section 4 includes a summary of the presented research and some proposals for future research directions.

5.2 The design process of collision avoidance (COLAV) algorithms for MASS

The process of ship collision avoidance is composed of environment perception, collision risk assessment, and the calculation of a collision avoidance action or actions by the COLAV algorithm [3–12]. The main challenges of this process are the limited time to make a decision, the presence of many static and dynamic obstacles in the surrounding environment, and the changes in motion parameters of encountered (target) ships. Below, the vital assumptions and constraints, that must be considered during the process of designing a COLAV algorithm for MASS, listed and briefly described.

DOI: 10.1201/9781003686088-5

5.2.1 Assumptions

Among the vital assumptions that must be considered during the development of a COLAV algorithm for MASS, the following issues should be regarded:

- the sources of input data and the data fusion method;
- the method of static and dynamic obstacles modeling;
- the assumed type of navigational situation to be solved – open or restricted waters;
- the assumed model of own ship and encountered (target) ships' motion;
- the method of ensuring COLREGs compliance of solutions;
- the method of ensuring a safe distance between own ship and target ships during maneuvering, for example, applying a ship domain;
- the assumed acceptable run time of the COLAV algorithm;
- the method of dealing with changing strategies of target ships;
- the assumption regarding the repeatability of solutions for the same input data.

5.2.2 Navigational environment modeling

In the navigational environment perception task, data from different sensors and systems, such as a log, a gyrocompass, the Global Navigation Satellite System (GNSS), a radar with the Automatic Radar Plotting Aid (ARPA), the Automatic Identification System (AIS), an echo sounder, and the Electronic Chart Display and Information System (ECDIS), are being registered and fed into the algorithm [13]. The following data constitute the input data to the COLAV algorithm:

- position coordinates (x, y), the course Ψ and the speed V of an own ship;
- courses Ψ_j and speeds V_j of target ships;
- bearings on target ships N_j;
- distances of target ships from an own ship D_j;
- positions of areas of land, shallows and other static obstacles.

The current navigational situation, as described by the data listed above, needs to be modeled in the COLAV algorithm. The navigational environment is defined as a two-dimensional space with boundaries consisting of a free area and an area occupied by obstacles. Two types of obstacles can be identified: static obstacles, like land, buoys, and shallows, and dynamic obstacles that change their position over time. Dynamic obstacles refer to ships encountered, also known as target ships. Therefore, the modeling of the navigational environment includes methods for representing of static and dynamic obstacles.

5.2.3 Consideration of obstacles

Dynamic obstacles (target ships) are commonly modeled by applying a ship domain around the current position of a ship. Different shapes and sizes of the ship domain can be applied, depending on the system operator's preferences [14,15].

Static obstacles are modeled using concave and convex polygons, representing the boundaries of an area occupied by obstacles or with the use of a cell decomposition method. In this approach the navigational environment is decomposed into free and occupied cells. An additional safety margin should be applied around the boundaries of static obstacles to ensure a safe distance between the own ship and static obstacles during the vessel's passage along a trajectory calculated by the COLAV algorithm.

5.2.4 Collision risk assessment

An important aspect of the COLAV approaches is the collision risk assessment. The safety measures used for that purpose are the DCPA – the Distance at the Closest Point of Approach, which is the minimum distance between an own ship and a target ship during both ships' movement along their trajectories, and the TCPA – the Time to the Closest Point of Approach. The TCPA is the time at which a target ship will reach its Closest Point of Approach. Other terms and measures used in the process of collision risk assessment include the ship domain, the degree of domain violation (DDV) and the time to domain violation (TDV), the Predicted Point of Collision (PPC), the Predicted Area of Danger (PAD), and the Collision Threat Parameters Area (CTPA) [16].

5.2.5 Solution construction methods

The solution construction procedure applied in the COLAV algorithm can be divided into two stages. In this approach, a set of feasible maneuvers or sets of maneuvers (safe trajectories) must be determined first. Subsequently, these candidate solutions are evaluated using an optimization criterion or criteria, to determine the most optimal trajectory, such as the shortest trajectory, as the best solution. An alternative approach could involve determining a feasible (safe) and the most optimal trajectory in a single stage.

Solution construction methods can also be divided into approaches based on the modification of an initial solution (trajectory) or a set of solutions in order to obtain a better solution in further iterations based on the initial one. Alternatively, they can be based on a constructive approach, where an algorithm, repeating a number of steps, builds a solution (a safe trajectory) starting from an empty solution [17].

5.2.6 COLREGs compliance of solutions

The COLAV algorithm should ensure the calculation of a safe maneuver or a set of maneuvers compliant with the International Regulations for Preventing Collisions at Sea (COLREGs). These are the navigation rules issued by the International Maritime Organization (IMO). The main rules related to COLAV are rules 8 and from 13 to 17 in Part B – Steering and Sailing Rules [18]. Rule 8 defines that the COLAV maneuver (course and/or speed alteration) should be large enough to be readily apparent to another vessel. Complementing this definition, it should be stated that it is undesirable to perform a sequence of small alterations of course and/or speed. Rules 13, 14, and 15 define actions to be taken by vessels in the following types of encounters: overtaking, head-on, and crossing. Most commonly applied methods of dealing with the consideration of COLREGs in the COLAV algorithm include the usage of if-then rules, an approach based on the determination of the vessels' priority, proper modeling of the ship domain shape and size, and the evaluation of the COLREGs compliance of the solution in the fitness function.

5.2.7 Evaluation of solutions

Solutions returned by the COLAV algorithm (safe single maneuvers or safe trajectories composed of a set of maneuvers) must be evaluated to verify their efficiency. The optimization criteria most used for that purpose are the path length (distance), the transition time, and the path smoothness.

The COLAV algorithm should also be evaluated considering the following features: the safety (feasibility) of returned solutions, the efficiency (optimality) of returned solutions, the reliability of the applied method, the repeatability of results for every run of calculations with the same input data and the run time of the algorithm (the calculation time), which preferably should be near real-time.

5.3 Results of simulation tests

This section presents examples of solutions obtained by five different COLAV algorithms [19,20].

5.3.1 A simple encounter situation

The input data for test case 1 is given in Table 5.1. The numerical results for this test case are listed in Table 5.2 and graphically in Figures 5.1 and 5.2.

Table 5.1 Input data for test case 1 with 1 target ship

Ship	*Ψ (°)*	*V (kn)*	*N (°)*	*D (nm)*
0	0	12	–	–
1	270	12	45	4

Table 5.2 Results of five different COLAV algorithms for test case 1

Algorithm	*Distance (nm)*	*Ψ (°)*	*Run time (s)*
ACO	9.22	11,346	5.55
TBA	9.22	14,349	0.11
DAPF	9.23	16,350	0.18
WAFR	9.23	16,350	0.46
VGA	9.06	10,355	0.69

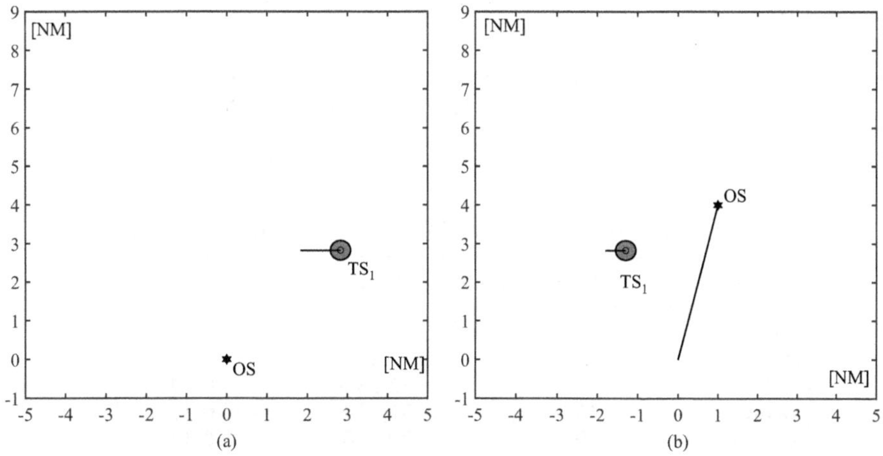

Figure 5.1 (a) initial configuration of test case 1 – OS at wp_s, (b) TBA solution of test case 1 – OS at wp_1.

5.3.2 *A multi-ship encounter situation*

The input data for test case 2 is given in Table 5.3. The numerical results for this test case are listed in Table 5.4 and graphically in Figures 5.3 and 5.4.

5.3.3 *An encounter situation with a static obstacle*

The input data for test case 3 is given in Table 5.5. The numerical results for this test case are listed in Table 5.6 and graphically in Figures 5.5 and 5.6.

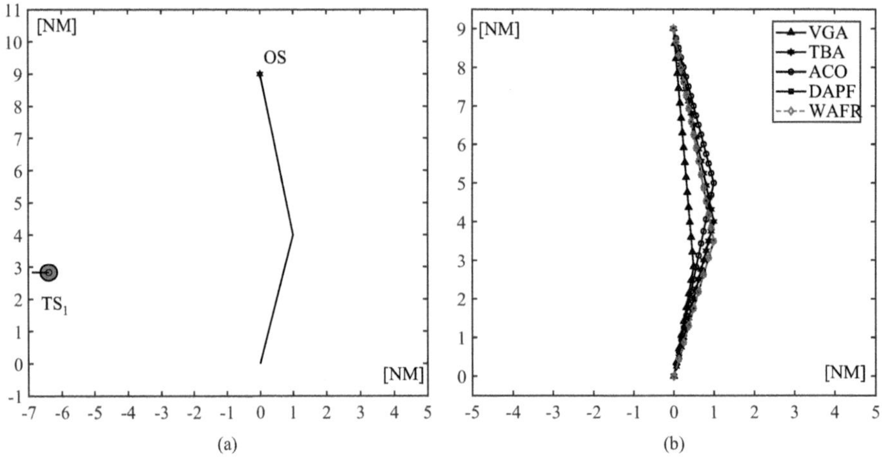

Figure 5.2 (a) TBA solution of test case 1 – OS at wp_g, (b) a comparison of safe paths calculated by five different COLAV algorithms for test case 1.

Table 5.3 Input data for test case 2 with 4 target ships

Ship	Ψ (°)	*V (kn)*	*N (°)*	*D (nm)*
0	0	14	–	–
1	270	14	45	4
2	200	2	20	8
3	175	10	345	7
4	270	14	55	6

Table 5.4 Results of five different COLAV algorithms for test case 2

Algorithm	*Distance (nm)*	Ψ (°)	*Run time (s)*
ACO	9.93	34,342	16.42
TBA	9.93	34,342	0.76
DAPF	10.31	32,333	0.19
WAFR	10.31	32,333	0.33
VGA	9.87	30,340	0.32

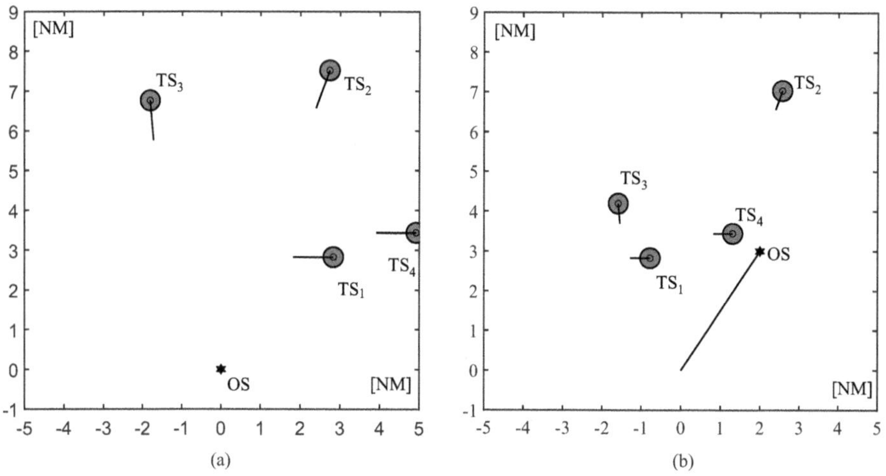

Figure 5.3 (a) initial configuration of test case 2 – OS at wp_s, (b) TBA solution of test case 2 – OS at wp_1.

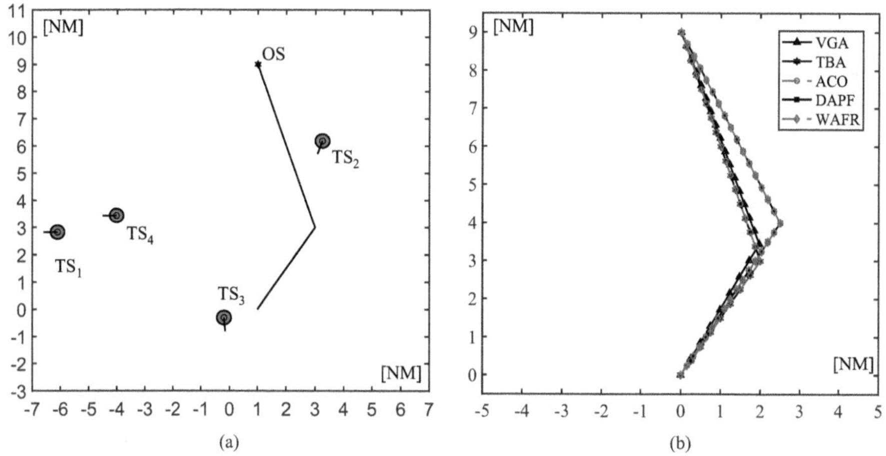

Figure 5.4 (a) TBA solution of test case 2 – OS at wp_g, (b) a comparison of safe paths calculated by five different COLAV algorithms for test case 2.

Table 5.5 Input data for test case 3 with 3 target ships and a static obstacle

Ship	Ψ *(°)*	*V (kn)*	*N (°)*	*D (nm)*
0	0	10	–	–
1	270	11	25	7
2	270	12	40	8
3	270	10	41	9

Table 5.6 Results of five different COLAV algorithms for test case 3

Algorithm	*Distance (nm)*	*Ψ (°)*	*Run time (s)*
ACO	9.25	14,0,346	55.48
TBA	9.25	14,0,346	0.18
DAPF	9.59	13,329	0.09
WAFR	9.59	13,329	0.08
VGA	9.4	18,351,342	0.74

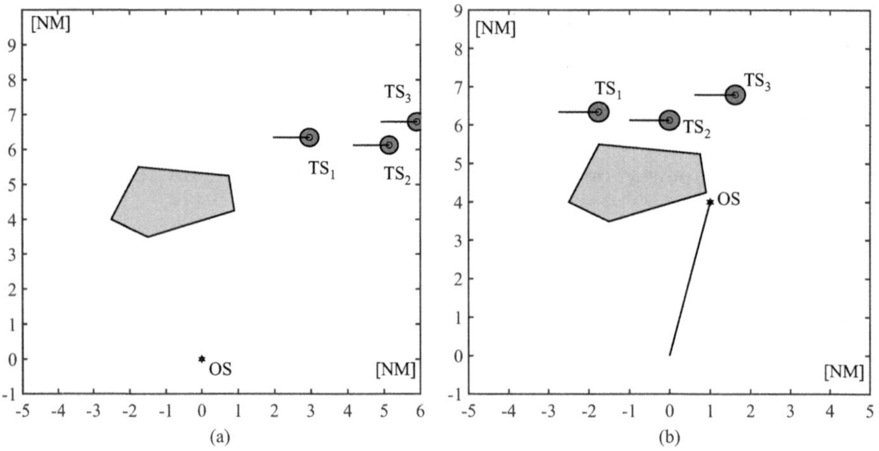

Figure 5.5 (a) initial configuration of test case 3 –OS at wp$_s$, (b) TBA solution of test case 3 -OS at wp$_1$.

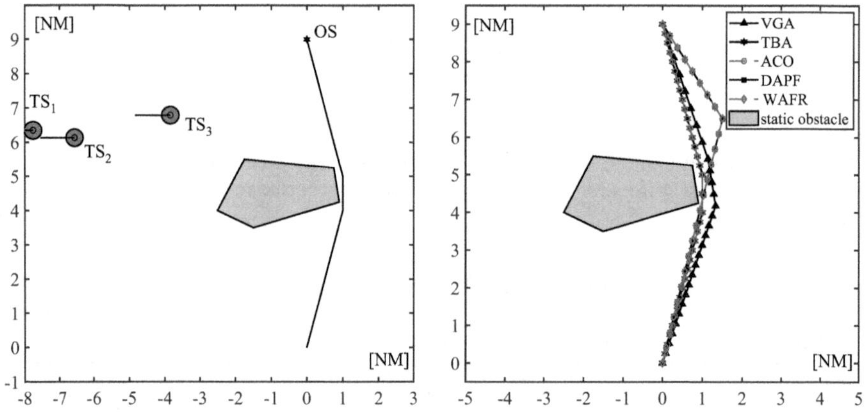

Figure 5.6 (a) TBA solution of test case 3 – OS at wp$_g$, (b) a comparison of safe paths calculated by five different COLAV algorithms for test case 3.

5.4 Conclusions

The most important assumptions, constraints, and procedures, that must be considered in the process of designing a COLAV algorithm for MASS or USVs have been described in this paper, along with the presentation of some examples of obtained results. The design process of a COLAV algorithm should include the following steps: the description of assumptions, the navigational environment modeling, the consideration of obstacles, the development of solution construction method, ensuring the solutions comply with COLREGs, and verifying the approach method. Future research directions defined by the author include: developing algorithms for determining a safe ship trajectory using machine learning methods, further testing the developed algorithms and verifying solutions in real or near-real conditions.

References

1. *Mayflower Autonomous Ship.* www.mayflower400uk.org/mayflower-autonomous-ship/.
2. *EHang AAV.* www.ehang.com/ehangaav/.
3. Lisowski J., 2024, Computational intelligence supporting the safe control of autonomous multi-objects, *Electronics*, Vol. 13(4), 780. https://doi.org/10.3390/electronics13040780
4. Liu Y., Bucknall R., 2015, Path planning algorithm for unmanned surface vehicle formations in a practical maritime environment, *Ocean Engineering*, Vol. 97, 126–144. https://doi.org/10.1016/j.oceaneng.2015.01.008
5. Łącki M., 2021, An adaptive island model of population for neuroevolutionary ship handling, *Polish Maritime Research*, Vol. 28(4), 142–150. https://doi.org/10.2478/pomr-2021-0056
6. Szłapczyński R., Szłapczyńska J., Gil M., Życzkowski M., Montewka J., 2024, Holistic collision avoidance decision support system for watchkeeping deck officers, *Reliability Engineering & System Safety*, Vol. 250, 110232. https://doi.org/10.1016/j.ress.2024.110232
7. Zaccone R., 2024, A dynamic programming approach to the collision avoidance of autonomous ships, *Mathematics*, Vol. 12(10), 1546. https://doi.org/10.3390/math12101546
8. Zhang K., Huang L., He Y., Wang B., Chen J., Tian Y., Zhao X., 2023, A real-time multi-ship collision avoidance decision-making system for autonomous ships considering ship motion uncertainty, *Ocean Engineering*, Vol. 278, 114205. https://doi.org/10.1016/j.oceaneng.2023.114205
9. Zheng K., Zhang X., Wang C., Li Y., Cui J., Jiang L., 2024, Adaptive collision avoidance decisions in autonomous ship encounter scenarios through rule-guided vision supervised learning, *Ocean Engineering*, Vol. 297, 117096. https://doi.org/10.1016/j.oceaneng.2024.117096
10. Zheng M., Xie S., Chu X., Zhu T., Tian G., 2020, Research on autonomous collision avoidance of merchant ship based on inverse reinforcement learning, *International Journal of Advanced Robotic Systems*, Vol. 17. https://doi.org/10.1177/1729881420969081
11. Zheng M., Zhang K., Han B., Lin B., Zhou H., Ding S., Zou T., Yang Y., 2024, An improved VO method for collision avoidance of ships in open sea, *Journal of Marine Science and Engineering*, Vol. 12(3), 402. https://doi.org/10.3390/jmse12030402
12. Zubowicz T., Armiński K., Witkowska A., Śmierzchalski R., 2019, Marine autonomous surface ship – control system configuration, *IFAC-PapersOnLine*, 8, 409–415. https://doi.org/10.1016/j.ifacol.2019.08.100
13. Bole A., Wall A., Norris A., 2014, *Radar and ARPA Manual*, Elsevier Butterworth-Heineman. https://doi.org/10.1016/C2010-0-68325-8
14. Szlapczynski R., Szlapczynska J., 2017, Review of ship safety domains: Models and applications, *Ocean Engineering*, Vol. 145, 277–289. https://doi.org/10.1016/j.oceaneng.2017.09.020
15. Zarzycki F., Gil M., Montewka J., Szłapczyński R., Szłapczyńska J., 2025, Declarative ship arenas under favourable conditions, *Ocean Engineering*, Vol. 316, 119927. https://doi.org/10.1016/j.oceaneng.2024.119927

16. Szlapczynski R., Szlapczynska J., 2016, An analysis of domain-based ship collision risk parameters, *Ocean Engineering*, Vol. 126, 47–56. https://doi.org/10.1016/j.oceaneng.2016.08.030
17. LaValle S.M., 2006, *Planning Algorithms*, Cambridge University Press, Cambridge.
18. Cockcroft A.N., Lameijer J.N.F., 2012, *A Guide to the Collision Avoidance Rules*, Elsevier Butterworth-Heineman. https://doi.org/10.1016/C2010-0-68322-2
19. Lazarowska A., 2022, *Safe Trajectory Planning for Maritime Surface Ships, Vol. 13, Springer Series on Naval Architecture, Marine Engineering, Shipbuilding and Shipping*, Springer Cham, Switzerland. https://doi.org/10.1007/978-3-030-97715-3
20. Rybczak M., Lazarowska A., 2024, Safe control of an autonomous ship in a port based on ant colony optimization and linear matrix inequalities, *Electronics*, Vol. 13(17), 3525. https://doi.org/10.3390/electronics13173525

Chapter 6

Supporting maritime autonomous systems with intelligent control methods

Józef Andrzej Lisowski

6.1 Introduction

The introduction of Maritime Autonomous Surface Ships (MASS) into service began in 2017. In 2022, according to Market Reports [1], the global market for autonomous ships was valued at USD 3.9 billion and is estimated to reach USD 8.2 billion by 2030. According to Rolls-Royce [2], the timeline for the stages in the development of MASS technology is as follows:

- 2020—remotely controlled local state with a reduced crew, with remote support and support for some operational functions;
- 2025—remotely controlled unmanned coastal ship;
- 2030—remotely controlled unmanned oceangoing ship;
- 2035—autonomous unmanned oceangoing ship.

It was indicated that the MASS design should include the following functions: risk management, risk root cause analysis, human integration, and identification of backup and critical system states [3]. The MASS system architecture should consist of the following elements: autonomous navigation architecture, decision-making and action system, situational awareness and related technologies, sensor fusion technology, collision avoidance, traffic control and path following, and berthing and unberthing subsystems [4].

A control system configuration ensuring the autonomous operation of MASS in various operational conditions was presented [5]. The improvement of MASS systems in terms of better navigation, predictive analytics to prevent accidents, and financial efficiency by reducing operating costs, lowering labor costs, better traffic management, and reducing the risk of loss of human life was proposed [6]. Further improvements were presented [7] in terms of supporting navigation monitoring and command and control in cooperation with land-based operations centers. Additionally, the design of an intelligent decision-making system based on an improved proximal policy optimization algorithm was proposed [8,9].

The importance of automation levels in autonomous ships can be demonstrated using a fuzzy inference system [10]. It is also possible to develop an autonomous collision avoidance system built around the perception-decision-execution sequence [11]. Intelligent algorithms with variable control strategy were also used in the design of the autonomous MASS system [12,13].

Integration of theoretical processes and Bayesian networks also enables the design of an autonomous navigation system within MASS [14]. However, in the implementation of individual MASS operations, a holistic understanding of human factors plays an important role [15].

DOI: 10.1201/9781003686088-6

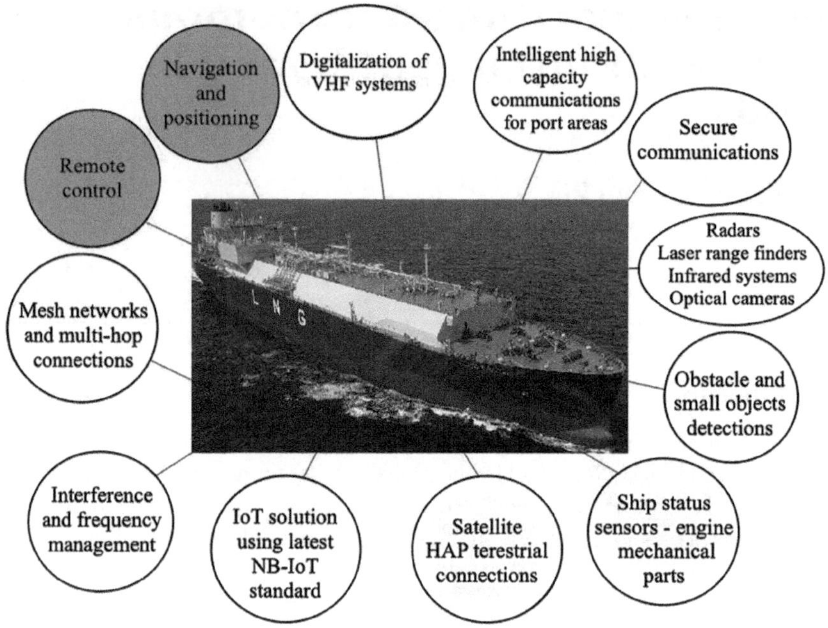

Figure 6.1 Functional diagram of an autonomous ship: VHF, Very High Frequency; HAP, High Altitude Platform; NB-IoT, Narrow Band-Internet of Things. (Freethink Media Inc., USA, 2025.)

The issue of MASS operation in accordance with the provisions of the COLREG convention is also important [16].

Figure 6.1 illustrates the functional scope of the autonomous MASS, with remote control, navigation, and positioning functions marked in green, which comprise the focus of this study.

Following a literature review on MASS control systems, this author formulated the intention for further research on the safe transport control process. The research aims to consider the subjectivity in making maneuvering decisions by applying game theory elements and address the risk of collisions in multi-object situations using a neural network. This study aims to create a synthesis of methods for controlling autonomous transport objects safely in collision situations caused by incorrect environmental assessments or subjective decision-making. The key innovation is integrating known methods of game theory, optimal control, and artificial intelligence to facilitate a better representation of the properties of an actual safe transport process.

6.2 Problem description

This work aims to select a safe and optimal control method for an autonomous object along a given motion trajectory. Various possible control approaches are available in a given situation, as shown in Figure 6.2. To achieve this optimization, we review various approaches supported by game theory and artificial intelligence, along with numerous optimization methods, considering the potential risk of collision.

The effectiveness of safe and optimal Q index control of an autonomous ship depends on the complexity and adequacy of the mathematical model describing the process (Figure 6.3).

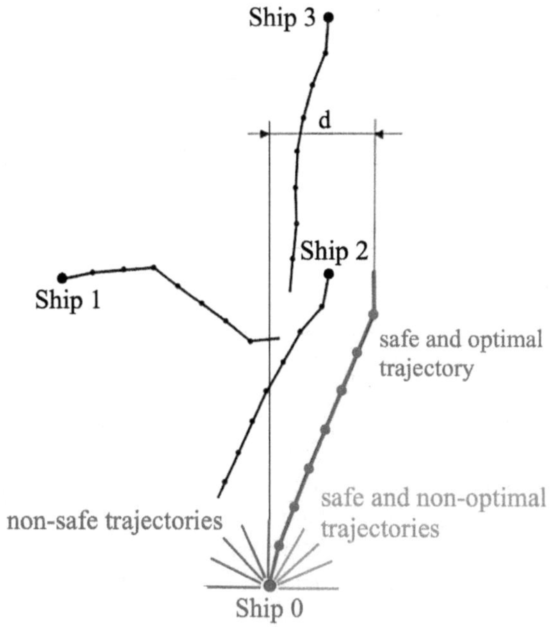

Figure 6.2 Possible controls for an autonomous ship.

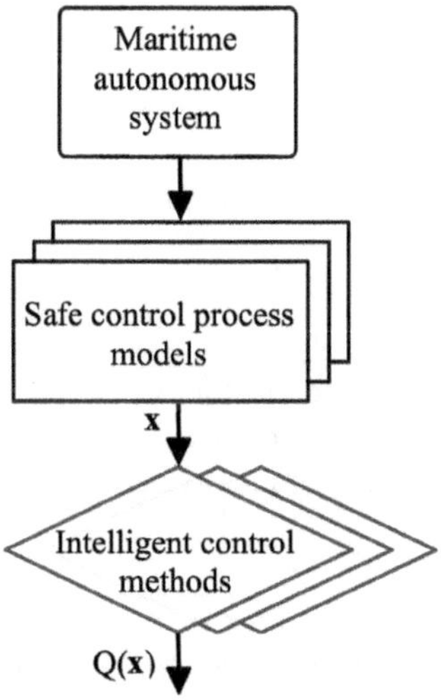

Figure 6.3 Autonomous maritime transport process control structure, where ***x*** comprises the state variables of the process control, and Q is the optimal control index.

The process for the safe control of an autonomous ship is described by the general state equation:

$$\dot{x}(t) = f(x, u, t) \tag{6.1}$$

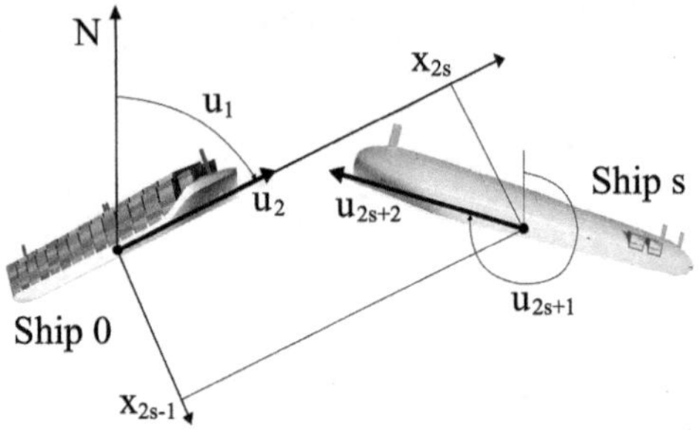

Figure 6.4 State variables describing the movement of autonomous ships: u_{2s+1} and u_{2s+2} are the course and speed of Ship s; u_1 and u_2 are the course and speed of Ship 0; x_{2s-1} and x_{2s} are the coordinates of Ship s.

where $\boldsymbol{x}$ represents state variables, $\boldsymbol{u}$ represents control varaiables, and $\boldsymbol{t}$ represents time.

The movement of autonomous ships, including the own ship (Ship 0) and another ship (Ship s), can be visualized in the geographic coordinate system shown in Figure 6.4.

The analysis of the relative motion of Ship 0 and Ship s yields the following equations for controlling the state of the autonomous ship as a differential game model:

$$\begin{aligned}
\dot{x}_1 &= u_2\left(ax_2 - 1\right) + u_4 \cos\left(\pi + u_1 - u_3\right) \\
\dot{x}_2 &= -ax_1u_2 + u_4 \sin\left(\pi - u_1 - u_3\right) \\
\dot{x}_3 &= u_2\left(ax_4 - 1\right) + u_6 \cos\left(\pi + u_1 - u_5\right) \\
\dot{x}_4 &= -ax_3u_2 + u_6 \sin\left(\pi - u_1 - u_5\right) \\
&\ldots\ldots\ldots\ldots\ldots\ldots \\
\dot{x}_{2s-1} &= u_2\left(ax_{2s} - 1\right) + u_{2s+2} \cos\left(\pi + u_1 - u_{2s+1}\right) \\
\dot{x}_{2S} &= -ax_{2s-1}u_2 + u_{2s+2} \sin\left(\pi - u_1 - u_{2s+1}\right) \\
s &= 1, 2, \ldots, S
\end{aligned} \tag{6.2}$$

where u_{2s+1} is the Ship s's course direction (deg), u_{2s+2} is its speed (m/s), x_{2s} is the current X-coordinate of Ship s (m), and x_{2s-1} is the current Y-coordinate of Ship s (m).

Optimal control should provide a minimum quality index:

$$Q(x) = \int f_o(x, u, t)dt \tag{6.3}$$

Furthermore, optimal control of an autonomous ship should ensure the least losses in safely passing encountered ships:

$$Q^*(t) = \int u_2 dt = u_2 t_0 \rightarrow \min \tag{6.4}$$

6.3 Methodologies

6.3.1 *Optimal game control*

The differential game is simplified to a kinematic multi-step optimal control game by replacing the description of the own object dynamics with the advance time of the course or speed change maneuver. Such a control task can then be solved using the triple linear programming method.

For cooperative optimal game control, the optimal control u_0^* of autonomous Ship 0 is determined using the following triple linear programming algorithm:

$$\min_{u_0^*\left(u_1^*,u_2^*\right)} \; \min_{u_s\left(u_{2s+1},u_{2s+2}\right)} \; \min_{u_{0,s}\left(u_{1s},u_{2s}\right)} t_0 = t_0^* \tag{6.5}$$

However, for non-cooperative optimal game control, the optimal control u_0^* of autonomous Ship 0 is defined using the following triple linear programming algorithm:

$$\min_{u_0^*\left(u_1^*,u_2^*\right)} \; \max_{u_s\left(u_{2s+1},u_{2s+2}\right)} \; \min_{u_{0,s}\left(u_{1s},u_{2s}\right)} t_0 = t_0^* \tag{6.6}$$

A saddle point in the cooperative or non-cooperative game is only possible when Ship 0 passes Ship s and only during one step of the trajectory. Based on Equation (6.2), which describes the kinematic motion of autonomous ships, and the principles of two-criteria game optimization (6.5) and (6.6), Algorithm 6.1 shows the pseudo-code for determining the optimal game control.

Algorithm 6.1: Optimal game control

Input: u_1, u_2, u_{2s+1}, u_{2s+2}, x_{2s-1}, x_{2s}
Output: X, Y, t

1. **begin**
2. step: $k = 1$
3. $k = k + 1$
4. autonomous ship: $s = 1$
5. $s = s + 1$
6. Ship 0 optimal control $u_{0s}\left(u_{1s}, u_{2s}\right)$ to Ship s
7. triple linear programming **GOTO** 8 and 9
8. cooperative game (5) **GOTO** 10
9. non-cooperative game (6) **GOTO** 11
10. Ship s cooperative control $u_s\left(u_{2s+1}, u_{2s+2}\right)$
11. Ship s non-cooperative control $u_s\left(u_{2s+1}, u_{2s+2}\right)$
12. **IF** not $s = S$ **THEN GOTO** 5
13. **ELSE** Ship 0 optimal control $u_0^*\left(u_1^*, u_2^*\right)$ to all Ships S
14. **IF** not $s = S$ **THEN GOTO** 3
15. **ELSE** safe optimal game trajectory of Ship 0
16. **end**

For the optimal game control algorithm simulation tests, an example navigation situation was adopted in which autonomous ships, Ship 0 and nine others (Ships 1–9), pass each other (Table 6.1).

Figure 6.5 shows the safe and optimal game trajectory of Ship 0 when cooperatively passing nine other autonomous ships. The safe and optimal trajectory of Ship 0 during the non-cooperative passing of nine other autonomous ships is shown in Figure 6.6.

Table 6.1 Navigational situation data of autonomous ship traffic

Ship s	*Course u_{2s+1} (deg)*	*Speed u_{2s+2} (m/s)*	*Coordinate x_{2s-1} (m)*	*Coordinate x_{2s} (m)*
0	0	11.0	0	0
1	291	4.0	−80	−415
2	44	5.0	305	−410
3	300	1.5	−445	120
4	0	4.0	690	425
5	90	8.0	−340	520
6	0	3.0	−425	685
7	340	8.0	−50	610
8	200	12.5	165	675
9	179	8.5	320	940

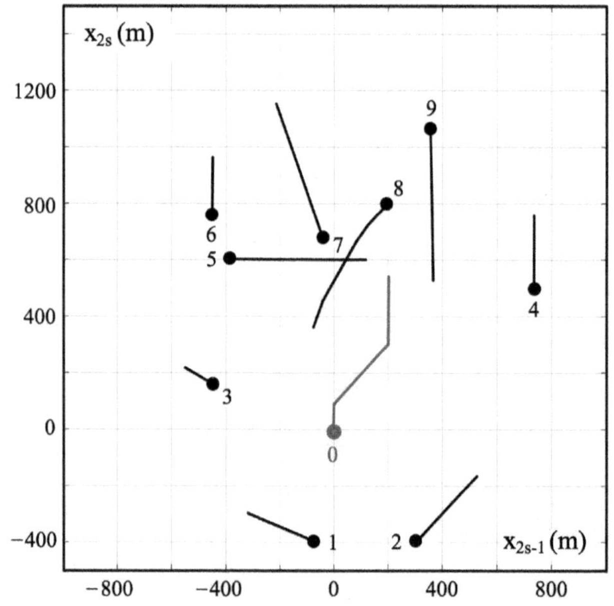

Figure 6.5 Optimal game trajectory of Ship 0 during safe cooperative passing of nine other autonomous ships.

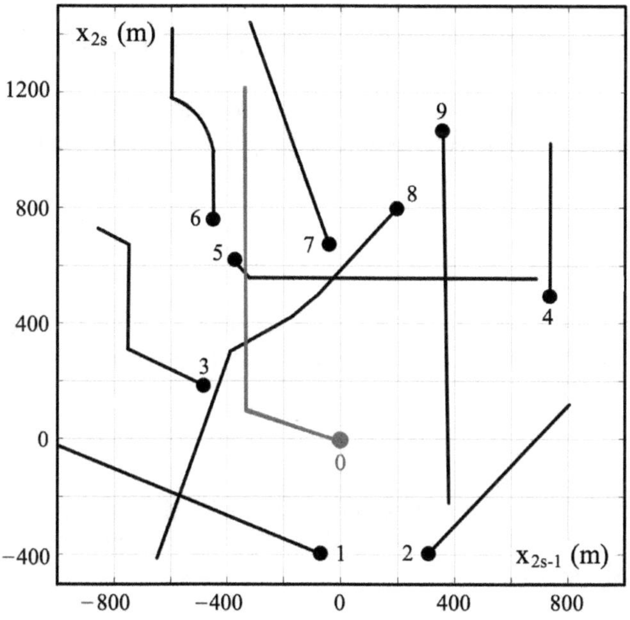

Figure 6.6 Optimal game trajectory of Ship 0 during safe non-cooperative passing of nine other autonomous ships.

6.3.2 *Risk game control*

The multi-stage risk game control is constructed as a matrix ***R*** containing the collision risk value r_s for *m* course-changing strategies of Ship 0 (rows) and *n* course-changing strategies of the Ships *s* (columns):

$$\mathrm{R}\left(r_{u_{1m}}, r_{u_{(2s+2)n}}\right) = \begin{bmatrix} r_{u_{11}u_{s1}}, \ldots, r_{u_{11}u_{sn}}, \ldots, r_{u_{11}u_{sN}} \\ \ldots\ldots\ldots \\ r_{u_{1m}u_{s1}}, \ldots, r_{u_{1m}u_{sn}}, \ldots, r_{u_{1m}u_{sN}} \\ \ldots\ldots\ldots \\ r_{u_{1M}u_{s1}}, \ldots, r_{u_{1M}u_{sn}}, \ldots, r_{u_{1M}u_{sN}} \end{bmatrix} \tag{6.7}$$

To calculate the collision risk value for an autonomous ship based on the measured state variables of the process, we employ an appropriate mathematical model:

$$r_s = \left(1 - 1.6 \log \frac{d_s^{\min}}{d_s}\right)\left(\frac{\dfrac{t_s^{\min}}{t_s} + 4}{8}\right) \tag{6.8}$$

where d_s and t_s are the given safe distance and time to the closest approach of Ship 0 and Ship s, respectively, and $d_s^{\min}$ and $t_s^{\min}$ are the minimum safe distance and time to the closest approach, respectively.

The risk game control problem can be reduced to dual linear programming as follows: Assuming the strategies of Ship 0 that minimize the risk of collision r_s and minimizing it via the strategies of each Ship s, a cooperative risk game control is obtained:

$$\min_{u_0} \min_{u_s} r_s = r_s^* \tag{6.9}$$

On the other hand, by using the strategies of Ship 0 to minimize the risk of collision and maximizing it through the strategies of each Ship s, a non-cooperative risk game control is achieved:

$$\min_{u_0} \max_{u_s} r_s = r_s^* \tag{6.10}$$

Based on Equation (6.2), which describes the kinematic motion of ships, and the principles of two-criteria game optimization (6.9) and (6.10), Algorithm 6.2 presents the pseudo-code for determining the risk game control.

Algorithm 6.2: Risk game control

Input: u_1, u_2, u_{2s+1}, u_{2s+2}, x_{2s-1}, x_{2s}
Output: X, Y, t

1. **begin**
2. step: $k = 1$
3. $k = k + 1$
4. autonomous ship: $s = 1$
5. $s = s + 1$
6. calculation of collision risk r_s (7)
7. determination of collision risk matrix $\boldsymbol{R}$ (6)
8. dual linear programming **GOTO** 9 and 10
9. cooperative game (9) **GOTO** 11
10. non-cooperative game (10) **GOTO** 12
11. **IF** not $s = S$ **THEN GOTO** 5
12. **ELSE IF** not $s = S$ **THEN GOTO** 3
13. **ELSE** safe optimal risk game trajectory of Ship 0
14. **end**

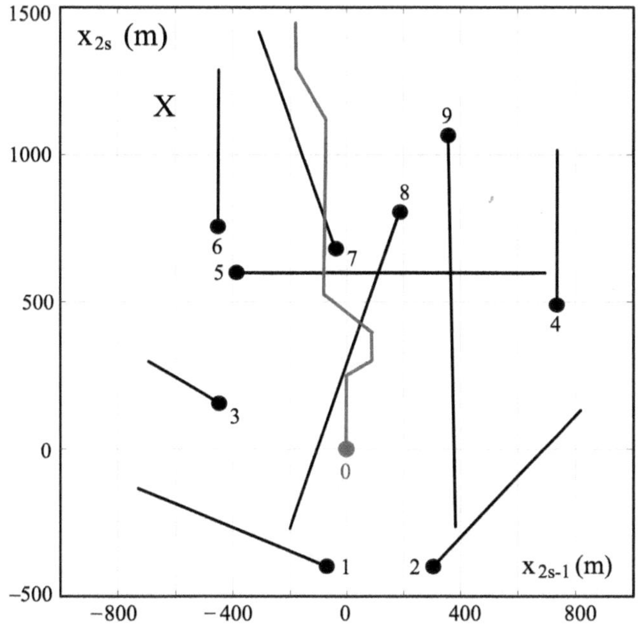

Figure 6.7 Risk game trajectory of Ship 0 during safe cooperative passing of nine other autonomous ships.

Figure 6.7 shows the safe and optimal game trajectory of Ship 0 in the situation described by the data in Table 6.1 for the cooperative passing of nine other autonomous ships according to the risk game control algorithm. The disadvantage of the risk game control algorithm is the inability to illustrate changes in the courses of other autonomous ships during computer simulation. The safe and optimal risk game trajectory of Ship 0 during the non-cooperative passing of nine other autonomous ships is shown in Figure 6.8.

6.3.3 *Neural dynamic control*

In neural dynamic control, when moving, autonomous ships are assigned state constraints in the form of forbidden motion domains generated by an artificial neural network. The constraints of the safe autonomous maritime transport process may take the following form: a parabolic area indicating excessive proximity of Ship 0 to Ship s:

$$C\left(x_{2s-1}, x_{2s}\right) = x_{2s-1} \sin u_{2s+1} + x_{2s} \cos u_{2s+1} - p\left(x_{2s-1} \cos u_{2s+1} + x_{2s} \sin u_{2s+1}\right)^2 \geq 0 \quad (6.11)$$

where p is the span of the parabola arms.

Such a neural network allows us to map the collision risk r_s and then adjust the size of the domain of prohibited maneuvers assigned to the passed Ship s, as shown in Figure 6.9.

Based on Equation (6.2) and the principle of optimization (6.11), Algorithm 6.3 is the pseudo-code for determining neural dynamic control.

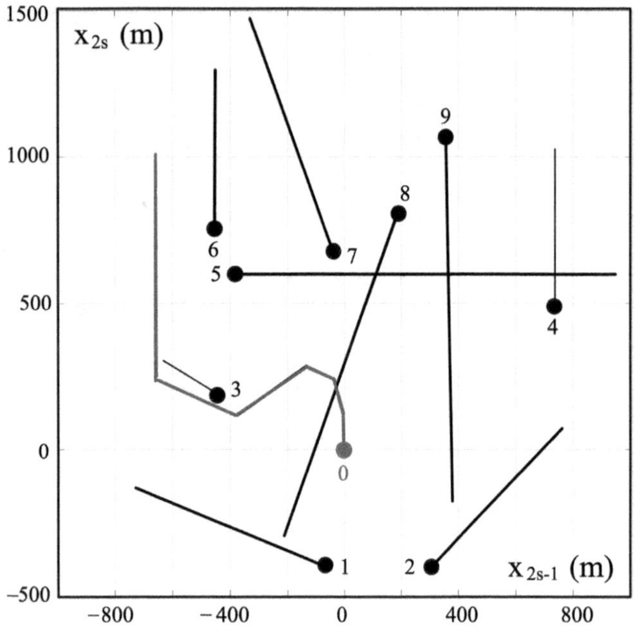

Figure 6.8 Risk game trajectory of Ship 0 during safe non-cooperative passing of nine other autonomous ships.

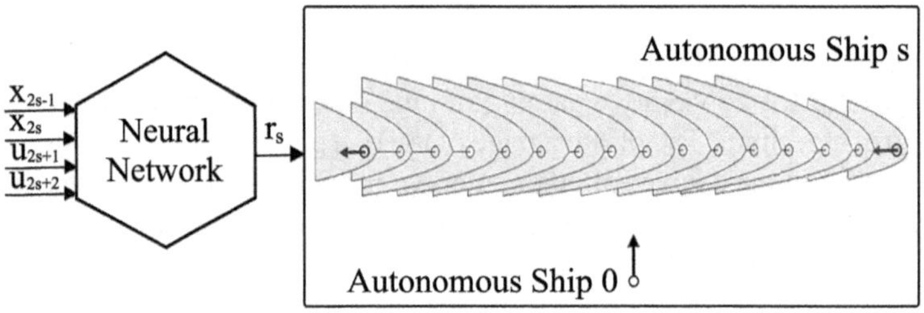

Figure 6.9 Generating constraints in the process for the safe control of autonomous ships using an artificial neural network.

Algorithm 6.3: Neural dynamic control

Input: u_1, u_2, u_{2s+1}, u_{2s+2}, x_{2s-1}, x_{2s}
Output: X, Y, t

1. **begin**
2. stage: $k=1$
3. $k=k+1$
4. node: $n=1$
5. $n=n+1$

6. autonomous ship: $s = 1$
7. $s = s + 1$
8. neural domain constraints (10)
9. **IF** not $k = K$ **THEN GOTO** 7
10. **ELSE** choosing optimal time path in Ship 0 in dynamic programming
11. **IF** not $n = N$ **THEN GOTO** 5
12. **ELSE IF** not $s = S$ **THEN GOTO** 3
13. **ELSE** safe optimal game trajectory of Ship 0
14. **end**

Figure 6.10 shows the safe and optimal trajectory of Ship 0 in the situation described by the data in Table 6.1, with nine other autonomous ships passing according to the neural dynamic control algorithm.

6.3.4 *Kinematic optimal control*

We assume that each encountered Ships s moves at a constant course u_{2s+1} and speed u_{2s+2}. The basis for the synthesis of control algorithms for multiple autonomous ships is first to establish sets of allowable course and speed changes that ensure a safe passing d_0 of Ship s (Figure 6.11).

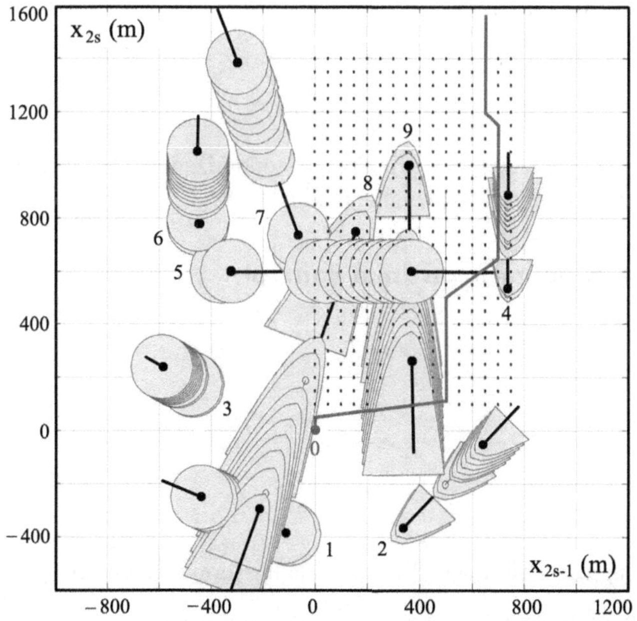

Figure 6.10 Neural dynamic trajectory of Ship 0 during the safe passing of nine other autonomous ships.

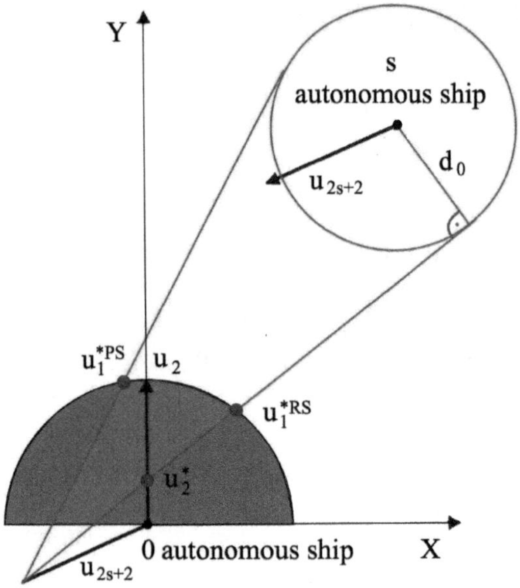

Figure 6.11. Determination of the sets of permissible maneuvers of Ship 0 (green), where optimal courses to the starboard and port sides are u_1^{*RS} and u_1^{*PS}, respectively, and the optimal maneuver occurs at a reduced speed u_2^*.

The optimal kinematic control $u_0^*\left(u_1^*, u_2^*\right)$ of Ship 0 is determined by the following single linear programming algorithm:

$$\min_{u_0^*\left(u_1^*, u_2^*\right)} t_0 = t_0^* \tag{6.12}$$

Based on Equation (6.2) and the principle of optimization (6.12), Algorithm 6.4 is the pseudo-code for determining the kinematic optimal control.

Algorithm 6.4: Kinematic optimal control

Input: u_1, u_2, u_{2s+1}, u_{2s+2}, x_{2s-1}, x_{2s}
Output: X, Y, t

1. **begin**
2. step: $k := 1$
3. autonomous ship: $s := 1$
4. determining permissible sets PS and RS in the form (Figure 6.11)
5. calculation of the optimal: courses u_1^{*RS}, u_1^{*PS} and speed u_2^* of Ship 0 according to criterion (12)

6. **IF** not $s = S$ **THEN** s := $s+1$ and **GOTO** 4

7. **ELSE** $k := K$

8. **IF not** $k = K$ **THEN** $k := k+1$ and **GOTO** 3

9. **ELSE** safe kinematic optimal trajectory of Ship 0

10. **end**

Figure 6.12 shows the safe and optimal trajectory of Ship 0 in the situation described by the data in Table 6.1, with nine other autonomous ships passing according to the kinematic optimal control algorithm.

6.4 Sensitivity control algorithm

To assess the quality of safe control of an autonomous ship, we choose to measure its sensitivity q to changes in the required safe distance d_0 of approaching autonomous ships, depending on traffic conditions and the assessment of the situation:

$$q = \frac{\partial d_f}{\partial d_0} = \frac{\frac{\Delta d_f}{d_f}}{\frac{\Delta d_0}{d_0}} = \frac{\Delta d_f}{\Delta d_0}\frac{d_0}{d_f} \tag{6.13}$$

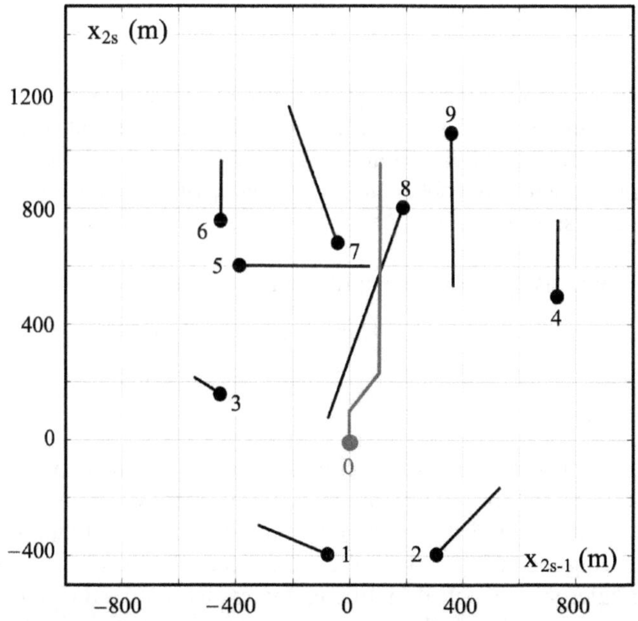

Figure 6.12 Kinematic optimal trajectory of Ship 0 during the safe passing of nine other autonomous ships.

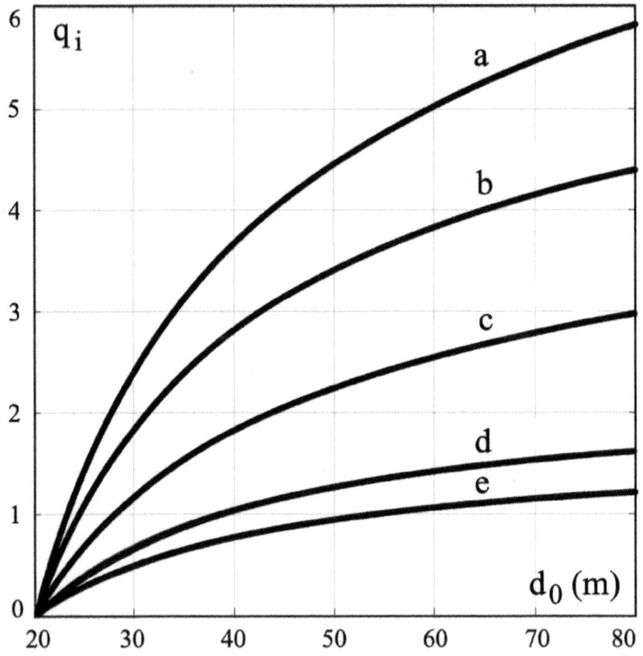

Figure 6.13 Characteristics for assessing the improvement in the safe control of autonomous ships using neural and game control methods of neural dynamic control (a), non-cooperative risk game control (b), non-cooperative optimal game control (c), cooperative risk game control (d), and cooperative optimal game control (e).

where d_f is the final deviation of the safe trajectory from its initial course.

By relating the sensitivity of the optimal game control, risk game control, and neural dynamic control methods to the sensitivity of the kinematic optimal control without considering artificial intelligence and game control, we obtain an assessment q_i of the relative sensitivity of the safe control of autonomous ships (Figure 6.13):

$$q_r = q\left(\text{optimal game, risk game, and neural dynamic control}\right) - q\left(\text{kinematic optimal control}\right) \tag{6.14}$$

6.5 Conclusions

The optimal game control and risk game control algorithms allow for cooperation or lack thereof during movement, providing a more accurate representation of real autonomous transport processes. The neural dynamic control algorithm is particularly sensitive to the risk of collision during the safe passage of an autonomous ship. The proposed relative sensitivity factor for safe object control facilitates the evaluation of the algorithm's effectiveness regarding safe control during the autonomous transport process and provides practical application support.

Further research expanding on this work may include the following aspects:

- Testing the sensitivity of safe control of multiple autonomous ships, considering inaccuracies in measured movement data and changes in dynamic properties caused by changes in load levels;
- Implementing the algorithms in programmable logic controller technology;
- Connecting the programmable logic controller to the ship's measuring devices;
- Developing a real-time depiction of the object's safe trajectory.

Acknowledgment

This study was financed by the research project "Development of methods and algorithms for environmental perception, navigation and control of autonomous vehicles" under grant number WI/2025/PZ/01 from the Faculty of Computer Science, Gdynia Maritime University, Poland.

References

1. Autonomous ships market. https://www.marketsandmarkets.com/Market Reports/autonomous-ships-market-267183224.html.
2. Autonomous ships – the next step. https://www.rolls-royce.com/~/media/Files/R/Rolls-Royce/documents.
3. Charalambous N. Risk based assessment tool for MASS (RBAT). *5th International Ship Autonomy and Sustainability Summit*, September 2024, Hamburg.
4. Alamoush AS, Olcer AI. Maritime autonomous surface ships: architecture for autonomous navigation systems. *Journal of Marine Science and Engineering*. 2025;13. https://doi.org/10.3390/jmse.2025.13010122
5. Zubowicz T, Armiński K, Witkowska A, et al. Marine autonomous surface ship – control system configuration. *IFAC-PapersOnLine*. 2019;52:409–415. https://doi.org/10.1016/j.ifacol.2019.08.100
6. IALA International Association of Marine Aids to Navigation and Lighthouse Authorities. *The Future of Maritime Autonomous Surface Ships (MASS), Future Scenarios Regarding the Development and Evolution of Mass*, 2024.
7. Guan W, Luo W, Cui Z. Intelligent decision-making system for multiple marine autonomous surface ships based on deep reinforcement learning. *Robotics and Autonomous Systems*. 2024;172. https://doi.org/10.1016/j.robot.2023.104587
8. Adnan M, Perera LP. Navigational support framework for maritime autonomous surface ships under onshore operation centers. *7th International Conference on Maritime Technology and Engineering MARTECH*, 2024, Lisbon, Portugal. https://doi.org/10.1201/9781003508779-41
9. Guan W, Cui Z, Zhang X. Intelligent smart marine autonomous surface ship decision. System based on improved PPO algorithm. *Sensors*. 2022;22:5732. https://doi.org/10.3390/s22155732
10. Poornikoo M, Overgard KI. Levels of automation in maritime autonomous surface ships (MASS): a fuzzy logic approach. *Maritime Economics & Logistics*. 2022;24(3). https://doi.org/10.1057/s41278-022-00215-z
11. Liu X, He Y, Zhang K, et al. Research on the decision-making method for autonomous navigation for the ocean-going ship in the ships' routing waters. *Ocean Engineering*. 2025;323:120641. https://doi.org/10.1016/j.oceaneng.2025.120641
12. Li S, Fung KS. Maritime autonomous surface ships (MASS): implementation and legal issues. *Maritime Business Review*. 2019;4(4):330–339. https://doi.org/10.1108/MABR-01-2019-0006
13. Wang L, Wu Q, Liu J, et al. State-of-the-art research on motion control of maritime autonomous surface ships. *Journal of Marine Science and Engineering*. 2019;7(12):438. https://doi.org/10.3390/jmse7120438

14. Laakso A, Chaal M, Valdez Banda OA. A risk assessment of an autonomous navigation system for a maritime autonomous surface ship. *Journal of Marine Engineering & Technology*. 2025; 2025:1–17. https://doi.org/ 10.1080/20464177.2025.2460268
15. Li X, Yuen KF. A human-centred review on maritime autonomous surfaces ships: impacts, responses, and future directions. *Transport Reviews*. 2024;44(4):791–810. https://doi.org/10.1080/01441647.2024.2325453
16. Lee HJ, Kim D. Navigating narrow waterways: a quantitative study on timing of stand-on ship actions for cooperative collision avoidance for MASS. *Ocean Engineering*. 2025;322:120400. https://doi.org/10.1016/j.oceaneng.2025.120400

Chapter 7

Managing the safe operation of ships in the context of the development of autonomous shipping

Alicja Mrozowska and Piotr Mrozowski

7.1 Introduction

Digitalization of the maritime industry has made it possible to transmit data at a distance and to combine telecommunications technology (connectivity), information technology (computerization), and mass media (means of broad-based social communication) using the electromagnetic field. In the current ongoing digital transformation, the maritime industry is taking advantage of technological development for more efficient and integrated shipping.

At the heart of the deep transformation currently underway, driven by technological innovation and automation, Maritime Autonomous Surface Ships (MASS) are on the horizon of technological transformation. It is a disruptive force that could redefine the management of ship operations, port-sea turnover, and the conduct of navigation in the modern world. The entry of ships with various levels of autonomy into service is a matter of time. In essence, autonomous ships are marine vessels capable of performing a wide range of tasks, functions, and assignments with a high level of autonomy, reducing or eliminating the need for direct human intervention in every element of the complex maritime transportation system. Modern technology has become a hallmark of modern times. Automation has reduced human errors in operating equipment and undoubtedly contributed to the development of the maritime industry, as it was intended to serve man. However, the dynamism of digital transformation has created a disharmony between technological development and the maritime safety environment. This situation is causing an intensification of efforts to analyzing the current safety situation. Despite a growing commitment to conducting analyses on the state of maritime safety, the matter has not been fully explored.

From this perspective, the current intensified work on the design, construction and deployment of MASS is concerning. Ongoing legislative efforts are aimed at classifying them as convention ships (subject to SOLAS'74) with various levels of autonomy (ranging from one to four). These MASS will navigate in maritime areas alongside conventional ships, and their passage routes can be managed and navigated them from anywhere in the world (a remote location outside the MASS). Therefore, an adequate level of safety during their operation on shipping routes carried out via both networks and systems in cyberspace and in actual operations at sea must be ensured before they embark on their first sea voyages.

It can be assumed that there is still time to develop solutions for the cybersecurity of this shipping innovation, but above all we need to focus on the ongoing safety needs of the emerging new virtual safety environment and current activities in which cross-border maritime trade is conducted.

The entry of MASS into operation will be possible when the relevant provisions of laws governing autonomous shipping are implemented. The International Maritime Organization announced at the 108th session of the Maritime Safety Committee (MSC) that MASS operations

DOI: 10.1201/9781003686088-7

will be regulated in 2032 by the Maritime Autonomous Surface Ship Code – MASS Code [1]. This means that MASS will be able to transport various types of goods by sea along the same sea routes as conventional ships. Intensive legislative work is being carried out at the IMO to prepare appropriate regulations compatible with the intensified new needs arising from digital transformation, including those affecting ongoing human activities at sea. An example of the IMO's work to date is the adoption of IMO Resolution 428(98) Maritime Cyber Risk Management in Safety Management Systems on June 16, 2017 [2,3]. The resolution identifies the need to manage cyber risks at sea and obliges shipowners to incorporate it into already established safety management systems defined by the International Safety Management Code – ISM Code.[1]

Another issue is the need to ensure that the entry of MASS into service and the development of autonomous shipping will result in a level of safety at sea that is not inferior to the one that has already been developed over decades. Therefore, it seems necessary to study and assess the level of safety and pay attention to the sensitive areas of the new type of shipping and establish a standard of safety management at sea. Moreover, it is necessary to provide a uniform standard for the preparation of requirements by shipowners/managers for the development of a safety management system (SMS). To this end, the provisions of the MASS Code and ISM Code were analyzed to propose the extension of the ISM Code and establish the shape of the requirements of SMS to ensure safe navigation of MASS and conventional ships on shipping routes. The proposed solutions are based on the authors' many years of experience in positions related to the implementation, maintenance, and improvement of the ISM Code in shipping companies. They are supported by expert interviews and lessons learned at MSC section meetings, as well as the needs arising from the development of autonomous shipping in the near future.

The purpose of the study is to propose systemic solutions that take into account sensitive areas of interference between vessels currently in operation and MASS, based on the provisions of the ISM Code, along with an extension of its provisions including safety management standards during autonomous transformation in transporting cargoes by sea.

The authors present the structure of the work, which will facilitate navigation through the study and allow readers to find the solution to the research problems: What components should the SMS include to ensure the safe operation of both MASS and conventional vessels? Moreover, it helps familiarize oneself with the most important issues related to the development of systemic solutions that can be implemented in the SMS for both conventional and MASS vessels by their owners or managers. The paper has been divided into sections, such as introduction, provisions of the ISM Code with the possibility of expanding to include autonomous shipping, provisions of the project for MASS Code development and gaps in ISM Code provisions, forming new pillars of maritime safety as MASS enters service building a safety management system for a conventional ship and the differences resulting from autonomy for MASS, the identified new functional requirements and differences – the shape of the management system, safety management standards in the context of autonomous shipping development, and conclusions.

The results of the research and analysis conducted will allow shipping company personnel to pay attention to emerging new challenges in the correlation between autonomous shipping and conventional shipping, as well as enable them to prepare for MASS operations. In addition, the proposed provisions also address the provision of cybersecurity and the need to implement the needs for maintaining a safety standard on both maritime transportation routes and in cyberspace.

Finally, future recommendations for the development of autonomous shipping are presented to address the clash between automated tradition and autonomous modernity.

7.2 The provisions of the ISM Code with the possibility of expanding in order to include autonomous shipping, based on functional requirements

The standard for managing the safe operation of ships was established by SOLAS`74, implemented in Chapter IX of the Convention [5]. The ISM Code's requirements are stringent, but versatile enough to allow shipowners to implement its provisions while taking into account the specifics of their operations and the types of ships/marine vessels they manage. The provisions of the ISM Code obligate the shipowner or the entity managing on its behalf to establish, implement, maintain, and improve a safety management system. SMS stands for "an organizational structure and documented system that enables the shipowner's personnel to effectively implement safety and environmental policies" [6].

A system that serves as a roadmap for ensuring the safe operation of a ship. It provides a structured and documented system that allows the shipowner's employees to effectively implement

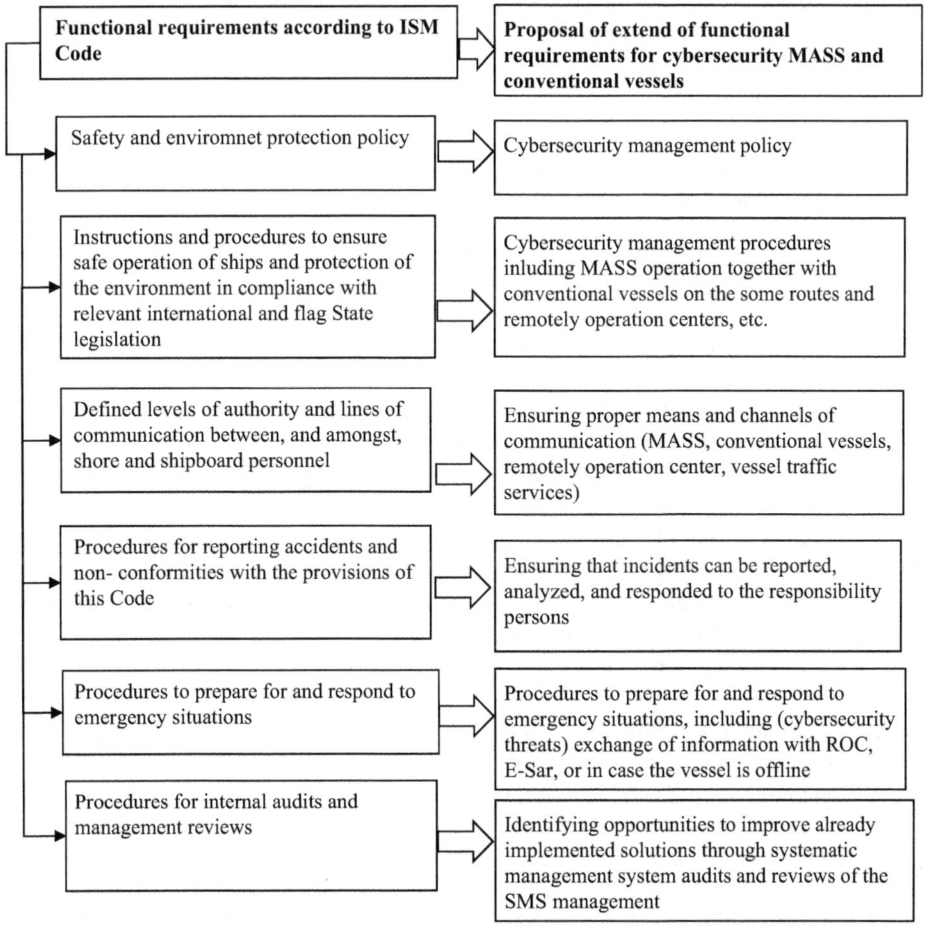

Figure 7.1 Functional requirements of the ISM Code and examples of their expansion, including cybersecurity during digital transformation and MASS operations. Prepared by the authors.

its safety and security policy. A properly functioning SMS is essential in eliminating risks to human life and health, material losses, and environmental pollution. Moreover, the shipowner/manager will not be allowed to operate ships if they do not meet the functional requirements.

Figure 7.1 presents the functional requirements of the ISM Code and examples of their implementation, including cybersecurity during digital transformation and MASS operations [1,7]. Therefore, the ISM Code appears to be an appropriate framework to complement the existing instruments issued by the IMO and provide regulations for the remote control and autonomous operation of key operations related to the safe operation of ships/marine vessels [8]. It serves as a means to compel both shipowners and ship managers to adopt a systemic approach to safety in both shipowners' offices and on ships/marine vessels.

MSC.428(98) resolution, as implemented in the ISM Code, which is Chapter IX of SOLAS`74, was intended to be the first step toward a systemic approach to ensure the appropriate level of cybersecurity for remote, autonomous ship operations with various levels of autonomy. Cybersecurity issues are not addressed in the MASS Code. Therefore, it seems that the ISM Code, together with the expanded IMO resolutions, will support the safety of MASS operations as well as conventional ships. Safety management systems will need to address individual issues of managing MASS operations and in correlation with MASS operations of conventional ships, as well as smart seaports.

7.3 The provisions of the project, the MASS Code development, and gaps in ISM Code provisions

The name "Maritime Autonomous Ships Surface" (MASS) was adopted at the 99th session of the MSC, a committee under the IMO [7].

According to the definition proposed by the IMO, a MASS means a ship that can operate at various levels independently of human intervention. MASS should be understood as ships with various levels of automation, ranging from ships with partially automated systems supporting crew operations to ships with fully autonomous systems capable of taking over all aspects of their operation without the need for human intervention [9].

The emerging new type of ship equipped with operational systems that allow remote control of its operation from a remote location outside the ship, or autonomous from human action to cover sea miles by means of algorithms, generates the challenge of implementing the appropriate level of safe operation of ships, taking into account the type and level of autonomy. In particular, international regulations must be established for these types of operations to ensure the level of safety at sea is not compromised.

There are four levels of automation:

1 **Level 1**: Ship equipped with automated decision support systems. Sailors are on board to operate and control ship systems and functions. Some operations may be automated and run unattended, but the sailors on board are ready to take control.
2 **Level 2**: Remote-controlled ship with sailors aboard. The ship is controlled and operated from another location. Sailors are aboard and can take control and operate shipboard systems and functions.
3 **Level 3**: A remote-controlled ship without sailors aboard. The ship is controlled and operated from another location.
4 **Level 4**: Fully autonomous ship. The ship's operating system is capable of making decisions and determining actions on its own.

In presenting the current distribution of MASS autonomy, it is important not to forget the previous automation in force on ships, which has become the nucleus for further expansion and development of autonomy in maritime transportation. For the purpose of maintaining a chronology of the development of autonomous navigation capabilities, the author introduced the concept of Level 0 autonomy. This term refers to the modern automation of conventional ships equipped with integrated systems used in the operation of the vessels.

According to the provisions of the MASS Code project, depending on the level of autonomy of the MASS, enforcement of safe navigation and environmental protection requirements is provided by the shipowner with the assistance of the crew and/or a possible remote crew.

Questions arise as to how the safe operation of MASS will be supervised in this not yet properly organized maritime virtual transportation world. To date, the IMO has proposed the following provisions in the draft 2023 MASS Code, based on the MASS oversight processes with the responsibility structure and functional requirements for MASS oversight. The MASS Manager (Shipowner) shall ensure continuous supervision and control of the MASS by the Captain and, if necessary, the crew aboard and/or remote off-board crew [10].

1 An unmanned (fully autonomous) MASS, excluding a fully autonomous passenger ship, equipped with autonomous systems that enable it to navigate from its point of departure to its destination and ensure safe navigation and protection of the marine environment without continuous supervision, is exempt from continuous supervision and control by a remote crew. MSC 107/20 report of the Maritime Safety Committee on its 107th Session, 26.06.2023.
2 MASS, excluding a fully autonomous ship, includes officers and operational personnel. MASS crews may combine the various functions specified in SOLAS`74 and the STCW`95 Code, provided that they maintain the established hours of work and rest and possess the qualifications required for each function performed.
3 A fully autonomous ship has no crew except for remote operation.
4 The remote crew provides remote control of MASS or assistance with MASS control to the crew aboard.
5 The remote crew may include remote MASS operators and responsible persons. Remote crew members carry out the shipowner's instructions regarding ship control, including ship navigation and work schedule. The charterer's instructions for the commercial operation of the ship are mandatory for remote crew members.
6 The captain (including the remote captain) is responsible for the MASS control, including navigation, taking measures to ensure the safety of the ship's navigation, protecting the marine environment, maintaining order aboard, preventing damage to the ship, as well as to people and cargo on board. The captain, including the remote captain, must hold a valid ship's captain's diploma and other certificates in accordance with the requirements of the applicable international instruments and national regulations established by the administration of the flag state.
7 In the case of the operation of a fully autonomous ship that does not have a remote crew, the shipowner must designate a person responsible for the management of the fully autonomous ship. This person will perform all functions and duties assigned to the ship's captain by applicable international instruments and the regulations of the flag state of the autonomous vessel.
8 The remote crew member directly controlling the MASS in the Remote Operations Center located off the ship is the remote operator. The flag state administration shall determine the requirements for vocational training programs for the operation of autonomous ships.

9 Remote operators and those in charge of the remote crew follow the instructions of the remote captain (MASS captain).
10 Operation of the MASS can be carried out by a remote crew that includes not only the shipowner but also an external organization that is competent in autonomous navigation and has the technical means to manage the MASS and meet the requirements established for remote operators. The flag state administration will determine the requirements for such organizations providing services for MASS operations.

Many questionable and unresolved issues remain, such as responding to states of emergency in which a fully autonomous ship will pose a threat to the marine environment, port, or sited infrastructure. This is no doubt an area for further research and analysis. The authors propose that emergency remote operation centers be included in the Code to operate in situations where the remote (operational) operation center is unable to take action, and it is necessary to support operations or immediately intervene in a critical situation during MASS operations. The immediate response will be based on the correct behavioral algorithm, but this case raises the need to take into account the ethics policy when preparing action algorithms for artificial intelligence. This task will be specified in the algorithms prepared by the operators. Making a decision by the responsible person for MASS operation and taking into account conventional vessels on shipping routes must be implemented according to international law pillars and then put into practice using safety management systems and cybersecurity prepared by the owner/ship management.

7.4 The formation of new pillars of maritime safety as MASS enters service, building a safety management system for a conventional ship, and the differences resulting from autonomy for MASS

A proposal to expand the modern pillars of maritime safety to include issues of autonomous navigation and safety standards during digital transformation at sea is shown in Figure 7.2. Accommodating the development of autonomous shipping with legal entrenchment is a challenge.

The proposed solution of expanding the provisions on autonomous navigation in the ISM Code has identified the areas that need to be considered when creating legal provisions. In order to ensure security standards on the vessel and in the ports, the ISPS Code was incorporated [11]. The provisions in a single document such as the draft MASS Code, do not address all the issues of legal compliance, especially those related to a systemic approach to ensuring shipping safety in maritime transport, the appropriate level of cybersecurity, and awareness arising from the resulting transition period between the operation of conventional ships and the intensification of the development of autonomy in shipping. Therefore, the prepared structure takes into account the need to implement harmonious provisions in the Codes: MASS, ISM, and ISPS. Another issue of direct conflict in the existing maritime legislation is the watchkeeping provisions of the STCW'95 Convention. With the assumptions of autonomous navigation, where the ships will be unmanned and operated from shore, existing requirements relating to the physical presence of a watch officer on the bridge are simply not going to be met. A more typical scenario is that numerous IMO regulations do not directly interfere with autonomous ships, but must be understood or interpreted in a particular way to enable autonomous or unmanned operations.

Such interpretations apply to all parties to the Convention and therefore cannot be made separately by individual parties (flag states). The current legal regime governing shipping is strongly anthropocentric and considers human control to be the most basic element of shipping safety.[2]

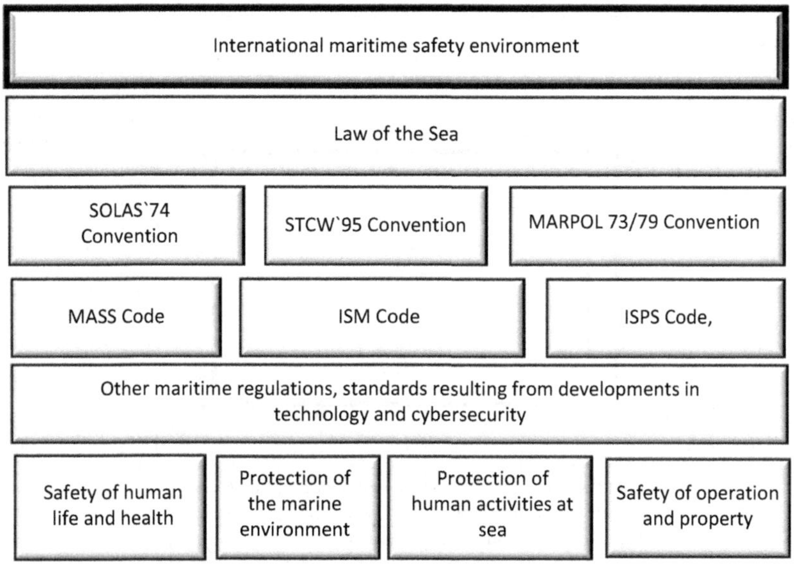

Figure 7.2 International maritime safety pillars expanded to include regulations in the context of autonomous shipping. Prepared by the authors.

Therefore, incorporating this issue into safety management systems established in accordance with the requirements of the ISM Code seems a possible direction for expanding the current pillars of maritime safety.

7.5 The factors influencing the structure of the safety management system, including cybersecurity management in the context of autonomous shipping development, and differences – stem and conclusions

Components of the MASS safety management system were proposed – taking into account the operation carried out by MASS, as shown in Figure 7.3.

Especially, the MASS safety management system is shaped by the following components:

- regulations established at the international and regional levels (international code conventions and other laws, MASS flag state regulations, classifier regulations, local-level regulations);
- MASS shipowner/manager or an organization recognized on its behalf (characteristics of the MASS shipowner/manager ☺
- MASS ship (type and level of MASS, OT and IT systems, cybersecurity of systems);
- area of operation of MASS;
- remote operation center (ROC) (functions, responsibilities, structure, competencies, etc.);
- cybersecurity management plan developed for MASS and ROC (information sharing and collaboration, authentication, and information safety attributes);
- Operations under the jurisdiction of the coastal state (seaport VTS surveillance area).

Additionally, the proposed provisions seem inadequate in a critical cybersecurity situation. The flag state should prepare a dedicated control center along the lines of the functioning Search and

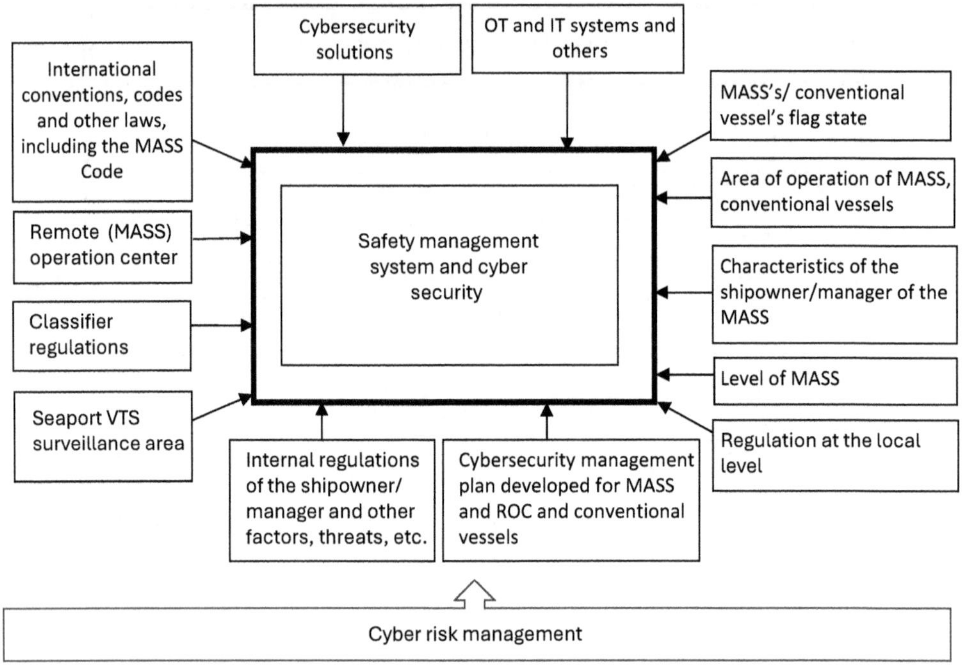

Figure 7.3 Elements influencing the design of the safety management system of MASS with various levels of autonomy. Prepared by the authors.

Rescue Service – SAR, the so-called cyberSAR. The assumptions adopted in the draft MASS Code form the nucleus for the organization of search and rescue operations by the relevant services, whose competence will lead to real time as well as cyber operations.

The foundation for establishing the system is to carry out cyber risk management This means that when implementing cyber risk management, safe operating practices in ship operations should identify risks and establish appropriate safeguards for the ship, personnel and environment within the ISM framework.

In summary, the future will not be a competition between the entities involved in providing maritime transportation, but ecosystems. Furthermore, due to technological developments, they will become common maritime cyberecosystems. Therefore, it is important to emphasize the need to intensify efforts to identify the different types of threats and risks including cyberspace, develop common mechanisms to cooperate, learn, and, above all, understand this need. This area will also need leaders, who know how to manage this ecosystem in a systemic way. Figure 7.4 depicts the assumptions for the purpose of utilitarianly preparing and then implementing a maritime transportation cybersystem with all its components and relationships occurring, allowing maritime transportation to be autonomous in its cybersecure environment.

Ports and ships are the hubs of the supply chain and global trade between maritime and land transport routes [12]. Given that most of the world's trade is conducted by sea, a major disruption at a strategic port or maritime bottleneck could have devastating consequences. With the adoption of digitization, most port and ship components use IT or OT systems in their core operations. These systems, previously unavailable, are increasingly using the port or ship's IT infrastructure to handle and monitor cargo turnover. The dissemination and analysis of data

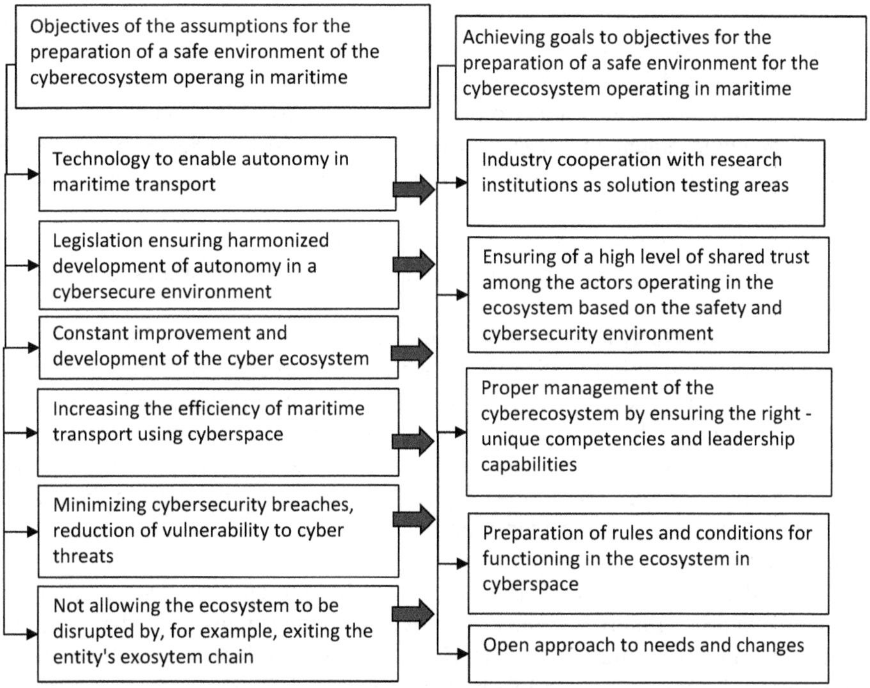

Figure 7.4 Assumptions for building a secure international maritime transportation cyberecosystem environment. Prepared by the authors.

collected from these components shape decisions. The correlations presented in the maritime transportation cyberecosystem lead to the conclusion that most components of both ecosystems, if affected by disruptions, can cause a physical domino effect. Maritime stakeholders should be proactive and develop business continuity mechanisms to respond effectively and in a timely manner in the event of a cyber-attack on any component of the MTS. States and relevant national authorities should encourage ship, port and terminal operators to act accordingly, while being mindful of relevant international laws and geopolitical tensions that may affect regional safety. An attack on a ship is not a remote scenario and will certainly affect the safe operation of the ship, navigation in the area as well as the operation of the seaport [13].

7.6 Conclusions

The intensified development of autonomous shipping and the deployment of MASS is only a matter of time. The MASS technology development being announced is highly advanced and ready for deployment. The ongoing tests based on regulations related to MASS sea trials and experiments at sea with autonomous shipping indicate that the lack of provisions in international law is preventing autonomous ships from entering service on international shipping routes.

Undoubtedly, a transitional period arises in various dimensions, during which a wise person should properly think, prepare and implement. Risk assessment, or rather risk-based thinking, becomes a tool. The purpose of the study was to propose systemic solutions that take into account sensitive areas of interference between vessels currently in operation and MASS, based on the provisions of the ISM Code, along with a proposed extension of its provisions. The proposed solution in the paper means:

First, it should be noted that in the proposed new structure of the international maritime safety environment, maritime safety continues to encompass the safety of life and health, as well as the protection of the environment and property exploited in maritime areas. Similarly, state maritime safety, also known as maritime security, refers to the protection of all human activities at sea. However, in the emerging maritime safety environment, they are taking on a new meaning formed by the digitalization of existing maritime activities.

Second, the functional requirements have been expanded to address issues related to ensuring cybersecurity management.

Third, due to cybersecurity issues not being addressed in the MASS Code, the MSC.428(98) resolution, as implemented in the ISM Code, which is Chapter IX of SOLAS`74, was to be the first step toward a systemic approach to ensure the appropriate level of cybersecurity for remote, autonomous ship operations with various levels of autonomy and concentrate on conducting proper cyber risk management. Thinking about risk helps to build a safety management system for a conventional ship and the differences resulting from autonomy for MASS, which is based on the identified new areas for safe operations at sea based on functional requirements.

Fourth, depending on the level of autonomy of the MASS, enforcement of safe navigation and environmental protection requirements is provided by the shipowner with the assistance of the crew and/or a possible remote crew. The MASS Manager (Shipowner) shall ensure continuous supervision and control of the MASS by the Captain and, if necessary, crew aboard and/or remote off-board crew. The rules and procedures must be established in SMS and cybersecurity.

The next step is that the MASS, conventional vessel, and remotely operated center for MASS should establish, implement, and ensure an actual SMS and cybersecurity.

Finally, the required quality of prediction will depend on the selected algorithms, considering factors such as the distance between the two ships, the safety margin each ship needs to maintain, real-time implementation of maneuvers to correct ship movements, limitations of the shipping basin, and the possibility of unexpected events occurring. When two ships are too close together and waters are restricted, it can be difficult or even impossible to implement a safe maneuver. The main prerequisite for safe interaction between a conventional ship and a MASS is to operate at the same operational level and exchange information effectively.

The era of digital transformation in shipping, where tradition meets modernity, is haggard with all sorts of disruptions. This situation creates a sense of unfamiliarity with the changes taking place, leading to uncertainty in a new, previously unexplored dimension. Therefore, the authors, in concluding their research, understand that they have not exhausted the topic, as there are not many papers on the subject, and hope that it will be further developed and continued by other researchers. In future research, it is important to consider the need to maintain a balance between the development of autonomy in shipping, innovation in technological progress and the safety environment, ethics, and progress. This includes taking into account the limits of human capability, working in extreme conditions, and the rational use of technology in an environment where nature sets its laws. It is also important to take into account the fact that humanity must continue to be wise and proactive rather than reactive and be able to skillfully use technological advances for the successful operation of maritime transportation in the maritime cyber domain.

Notes

1 Cyber threat means any potential circumstance, event or action that may cause damage, disruption or otherwise adversely affect networks and information systems, users of such systems and other persons. This definition is taken from Article 2 clause 8 [4].

2 Osaloni, O. S., Ayeni, V. O. The Development of Maritime Autonomous Surface Ships: Regulatory Challenges and the Way Forward, Beijing Law Review, Vol. 13, 544–554, 2022, https://www.scirp.org/journal/blr ISSN Online: 2159–4635, p. 550.

References

1. Bulletin PRS, *The Most Important Issues from 108 IMO MSC, 08/2024*, PRS, Gdansk, 2024.
2. IMO, *MSC-FAL.1/Circ.3/Rev.2 – Guidelines on Maritime Cyber Risk Management*, issue. 7/06/2022, London, 2022.
3. IMO, *MSC.428(98) – Maritime Cyber Risk Management in Safety Management Systems*, London, 2017.
4. The European Parliament and of the Council of 17 April 2019 on ENISA (the European Union Agency for Cybersecurity) and on information and communications technology cybersecurity certification and repealing Regulation (EU) No 526/2013 (Cybersecurity Act) L 151/15, 7.06.2019.
5. IMO, *International Management Code for the Safe Operation of Ships and for Pollution Prevention, ISM Code with Guidelines for its Implementation*, 2018 edition, London.
6. Miler, R.K. *Safety Sea Transport*, PWN, Warsaw, 2015.
7. IMO, *Draft MASS Code, latest version document MSC 109-5 (2024)*. https://docs.imo.org/[28.12.2024]
8. Mrozowska, A., Wrobel, R., Mrozowski, P. *System zarządzania bezpieczeństwem w zarysie – analiza, organizacja i metodyka oceny bezpieczeństwa*, Wydawnictwo Difin, Warsaw, 2022.
9. Bratić, K., Pavić, I., Vukša, S., Stazić, L. Review of Autonomous and Remotely Controlled Ships in Maritime Sector. *Transactions on Maritime Science*, No. 8(2), 253–265, 2019 https://ojs.ebujournals.lu/index.php/JIDS/article/view/18.
10. Osaloni, O.S., Ayeni, V.O. The Development of Maritime Autonomous Surface Ships: Regulatory Challenges and the Way Forward. *Beijing Law Review*, Vol. 13, No. 3, 544–554, 2022, https://www.scirp.org/journal/blr.
11. IMO, *Guide to Maritime Security and the ISPS Code*, 2021.
12. Wrobel, K., Montewka J., Kujala, P. Towards the Development of a System- Ceoretic Model for Safety Assessment of Autonomous Merchant Vessels, *Reliability Engineering & System Safety* 2018, Vol. 178, 209–224, 2018.
13. Meland, P.H., Bernsmed, K., Wille, E., Rodseth, O.J., Nesheim, D.A. A Retrospective Analysis of Maritime Cyber Security Incidents, *International Journal on Marine Navigation and Safety of Sea Transportation*, Vol. 15, No. 3, 2021. https://doi.org/10.12716/1001.15.03.04

Chapter 8

PSO method for optimizing loss function weights for ship model control

Monika Rybczak

8.1 Introduction

For several years, it has been observed that some swarm intelligence algorithms are utilized in research for path planning in various environments [1]. For example, genetic algorithms (GA) [2], particle swarm optimization (PSO) [3], or ant colony optimization (ACO) [4]. The first scientific publication on PSO appeared in 1995 [5], in which the authors referred to the use of algorithms for optimization based on PSO. Later, a book was published defining mathematical relationships for optimization, including multi-criteria tasks [6]. The work [7] in Figure 8.1 proposed a classification breakdown of the basic methods used in computational intelligence for neural networks, fuzzy logic, evolutionary computation, swarm intelligence, machine learning, and probabilistic reasoning. The PSO method is included in swarm intelligence. The PSO algorithm was first introduced by researchers in 2004 [8], demonstrating that the search for food by birds can be incorporated as an objective in the PSO method. It was shown that the search for food (a single piece) in a specific area is achievable by the bird closest to the food. The authors of the article [9] proposed a comprehensive review of literature related to the extensive modification of the PSO method at the turn of the millennium: from the classic version, through topologies and parameters, to hybridizations and applications.

The PSO method has found its application in the maritime industry most often for planning the trajectory of a ship. An interesting overview has been proposed by the authors [10–12], referring to navigation tasks, i.e., determining the safe trajectory of a ship based on the computational method in search of a global minimum for nonlinear optimization issues. Increasingly, researchers are applying PSO to collision avoidance scenarios for several ship objects [13,14].

In 2018, the IMO Convention introduced four states of autonomy related to autonomous control of a vessel [15,16]. Currently, the regulations are at the stage of introducing accuracy over the degrees of autonomy of a vessel and regulations related to COLREG regulations, among others. The paper [17] suggests future research on the accuracy and variety of development and application of MASS motion control based on methods of both classical regulation, computational intelligence methods and deep machine learning. This includes the adaptation of artificial intelligence models based on the relationship of environmental influences on the object.

Scientists around the world at the turn of the decade are increasingly turning to computational intelligence based on nature observation and have successfully implemented this in ship trajectory studies through DP-resistant control or autonomous unit control, among others. The author's considerations focus on a proposed control algorithm based on the computational intelligence method PSO.

DOI: 10.1201/9781003686088-8

The paper introduces an algorithm to search for the global minimum in the objective function, which impacts the synthesis of the state controller for the motion control of the ship during the maneuver of the training model in the port. The problem of finding a global minimum affects the operation of the linear matrix inequalities (LMIs), multivariable controller synthesis scheme. The proposed PSO method enables the optimization of a target function consisting of searching for the global minimum for two functions of norm H_∞ and norm H_2. First, a chapter describing the control object, which is a nonlinear training model of the Blue Lady ship, is presented. A maneuver performed in real conditions in the port of the Foundation for Safety of Navigation and Environmental Protection in Iława Kamionka is presented [18]. Next, a theoretical outline of the PSO method is presented, and a decision is made to optimize the ship's motion calculated according to the 3DOF Cartesian system. After analyzing the results, a decision was made to implement the PSO method.

8.2 Test site and control object

The research was carried out on Silm Lake in Kamionka at the Foundation for Safety of Navigation and Environmental Protection. The center is the only training site for seafarers in Poland. Training in ship handling is recommended by the IMO as one of the most effective methods of improving safety at sea. The purpose of the training conducted at the Foundation is to gain theoretical and practical knowledge of how to operate a ship in various situations encountered in practice at sea. Figures 8.1 and 8.2 show two shots of a section of the Foundation with the infrastructure prepared, a top shot (Figure 8.1) and a side shot (Figure 8.2).

8.2.1 Control object, model of training ship "Blue Lady"

There are nine models of surface ships in the Foundation for Safety of Navigation and Environmental Protection. Below are the contents of the table with model names and parameters, including actual dimensions and model dimensions. Most of the models were made in a 1:24 scale. Table 8.1 shows the dimensions of the ships in two rows and the names of the model units in nine columns.

Figure 8.1 "Overhead" shot of a view of the harbor on Lake Silm in Kamionka, near Ilawa, Poland [19].

Figure 8.2 Side view of the port located on Lake Silm in Kamionka, near Ilawa, Poland [18,20].

Table 8.1 Names and dimensions of real ships and their counterparts, training models.

No.	*Model's type*	*LCC tanker*	*Car carrier*	*Panamax bulk carrier*	*Ro-Ro vessel*	*Passenger car ferry*	*VLCC tanker*	*VLCS*	*ULCS*	*LNG carrier*
1	Length O.A.									
	Ship real [m].	292.93	218.85	227.75	199.90	175.48	330.65	372	394.4	277.2
	Model [m].	12.21	9.12	9.49	8.31	10.96	13.76	15.5	16.43	11.55
2	Breadth									
	Ship real [m].	48	30.5	32.1	31	28.5	57	43	53.6	43.2
	Model [m].	2	1.27	1.34	1.29	1.78	2.38	1.79	2.23	1.8

The research was performed on the VLCC training tanker "Blue Lady," whose vessel at the quay is shown in Figure 8.3. The ship is equipped with five propellers and a main propulsion system. More information can be found in [21,22].

A maneuver of the VLCC Tanker training vessel "Blue Lady" in the harbor was performed (Figure 8.4). A real-world maneuver was executed based on the multidimensional control of a nonlinear ship. The test site and the facility have been described in previous works [21–30]. The ship followed the six coordinate points marked in blue (points three and four are the same for better readability on the figure as one dot). The red trajectory illustrates the path followed by the autonomous system using multidimensional control, including the LMI controller and overriding trajectory, without PSO optimization.

The maneuver was performed for a total of 1300 seconds, which is more than 40 minutes. The maneuver took place in real conditions on Lake Silm. In Figure 8.5, the sequential readout

Figure 8.3 Ship handling of Blue Lady, L: 13.76 m, W: 2.38 m. It is equipped with two thrusters (stern and bow) and two tunnel thrusters (stern and bow), a rudder fin, and main propulsion.

values of the individual set and controlled speeds, wind speed, and DGPS system mode indications are presented.

The author's objective was to optimize the execution of the maneuver, leading to the decision to apply the PSO method. Specifically, the docking maneuver to the ALPHA 2 quay and longitudinal speed equalization were carried out over an extended period. The ship sailed backwards for nearly 300 seconds, recorded between 900 and 1200 seconds.

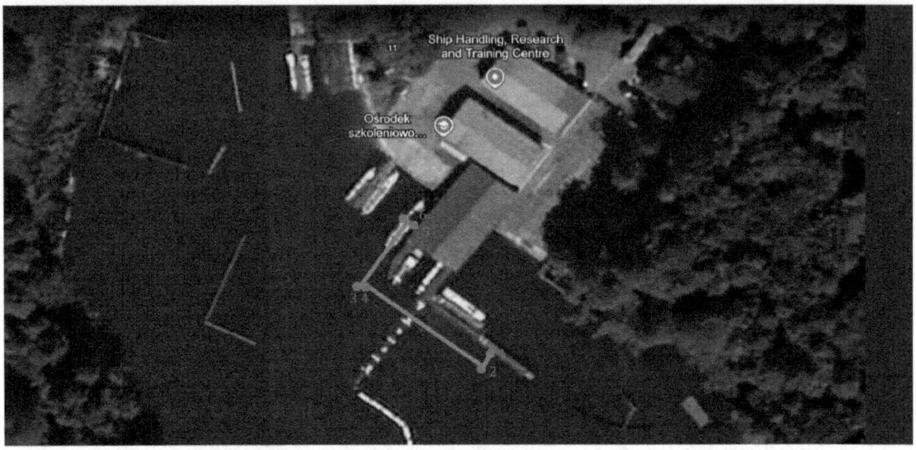

Figure 8.4 Coordinate values – points marked in blue, the route the ship traveled is marked in red. ALPHA 1 and ALPHA 2 are quay names.

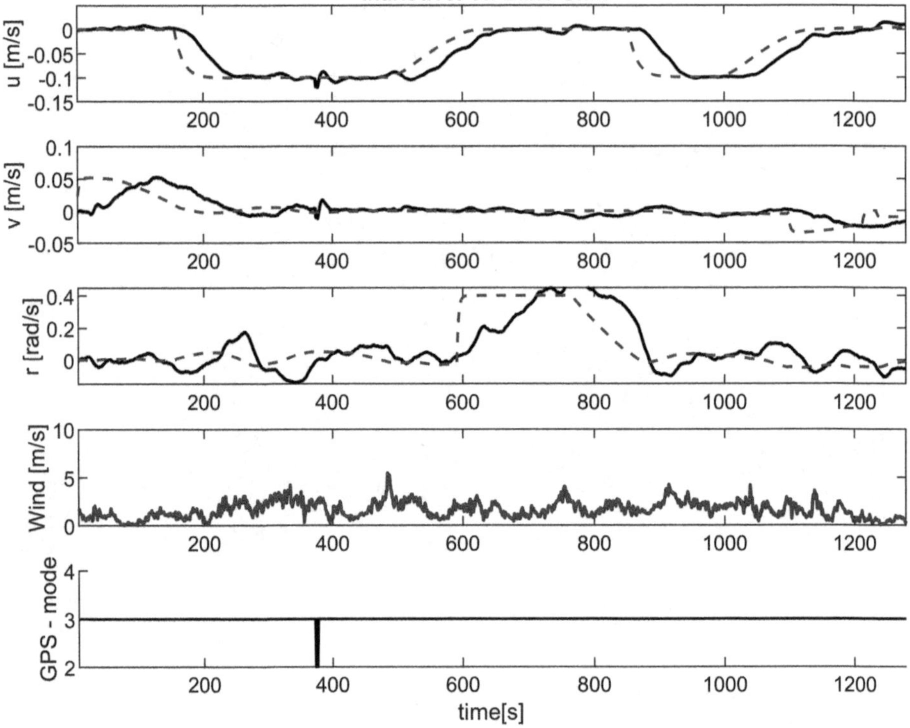

Figure 8.5 shows the recorded velocities during the execution of the test. The three set velocities are indicated by the blue dotted line, and the adjusted values are indicated by the black solid line. The recorded velocities are sequentially the longitudinal velocity denoted as u [m/s], the transverse velocity denoted as v [m/s], and the angular velocity denoted as r [rad/s].

8.2.2 *Autonomous control of the surface unit*

As the autonomous control system has already been extensively addressed in works [24,28], and the numerical methods involved are too complex to be fully covered in a single paper, the author limits the discussion to the main goals of synthesizing the two controllers. The first master controller, called the "Overriding trajectory controller," was concerned with establishing setpoints such as *x*, *y*, and heading coordinates based on the values of the three speed setpoints, which are necessary for further regulation when maneuvering in the port(Figure 8.6).

The calculations used are mainly concerned with the relationship of angles and path lengths for the ship, with the author of this work [33]. Professor Gierusz writes about the trajectory location recognition into three zones: zone one when the ship has not yet reached the expected trajectory, zone two when the ship is set to maximum values for the speed setpoints, and zone three when all settings are set to minimum. This process is shown in Figure 8.7.

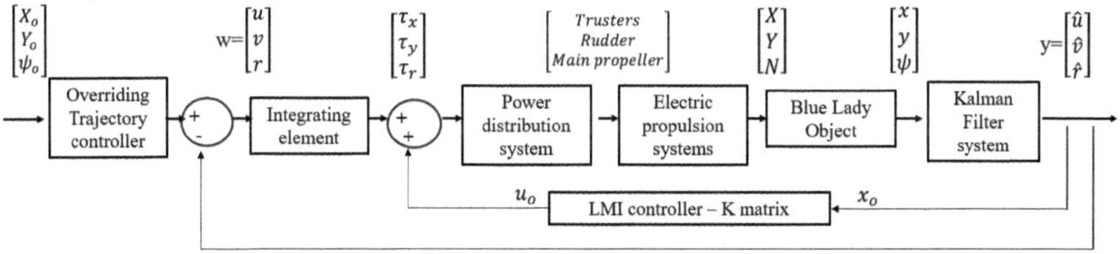

Figure 8.6 General schematic of a ship control system with overriding controller, overriding trajectory controller, and LMI controller [31,32]. Symbols u_0 and x_0 should be checked in Equation (8.2).

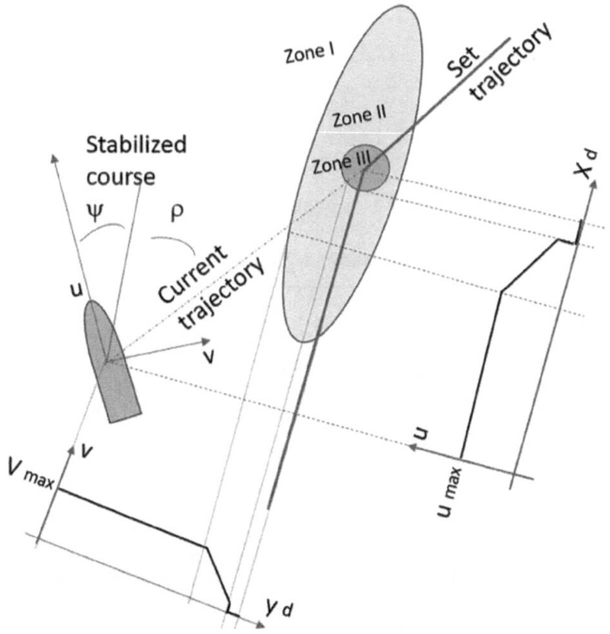

Figure 8.7 Graphic interpretation of controller operation [22,33].

The second state-to-state regulator is based on the methods of LMIs. As a result, the coefficients for the H_∞ and H_2 norms must be determined. Hence, the output signals for the state-to-state regulator have outputs that specify the signals needed to satisfy the individual y_∞ and y_2 norms and output the y signal (Figure 8.8).

Hence, the state matrix for the object (training state) is of the form Equation (8.1) (Table 8.2):

$$\begin{cases} \dot{x}_o = \left(A + B_u K\right)x_o + B_w w \\ y = \left(C + D_{yu} K\right)x_o + D_{yw} w \\ y_\infty = \left(C_\infty + D_{\infty u} K\right)x_o + D_{\infty y} w \\ y_2 = \left(C_2 + D_{2u} K\right)x_o + D_{2w} w \end{cases}, x_o = \begin{bmatrix} x_i \\ x \end{bmatrix} \tag{8.1}$$

The algorithm for controlling the controller matrix from the state involves searching for a positively specified matrix X, which results in the controller matrix K. In the simplified equation presented below, you can see Equation (8.2):

$$u_o = K \cdot x_o \tag{8.2}$$

Based on the Yalmip and SeDuMi 1.3 library in the Matlab 2023 environment, the K regulator matrix was obtained in the following form:

$$\mathbf{K} = \mathbf{10^3} \begin{bmatrix} \mathbf{1}.5690 & \mathbf{-0}.5490 & \mathbf{-0.0}973 & -0.7106 & \mathbf{0}.1843 & \mathbf{0}.0350 \\ \mathbf{-0.00}14 & \mathbf{1}.0181 & 0.0058 & \mathbf{-0.00}82 & -0.5908 & -0.\mathbf{00}54 \\ -0.1143 & 0.0873 & 0.2834 & 0.0360 & -0.0286 & -0.1597 \end{bmatrix}$$

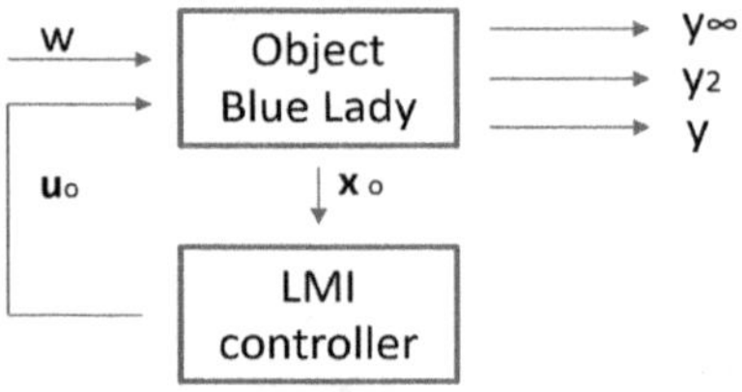

Figure 8.8 Graphic interpretation of the operation of the LMI slave controller.

Table 8.2 Matrix and name signals for system control

Matrix name	*Name signal*
A	Controlled object state space matrix
B_u	Control matrix of control signal "u";
B_w	Control matrix of input signal "w";
C	Output matrix of output signal "y";
C_∞	Output matrix of "y∞" signal;
C_2	Output matrix of "y2" signal;
D_{yu}	Transition matrix of "y" and "u" signals;
D_{yw}	Transition matrix of "y" and "w" signals;
$D_{\infty u}$	Transition matrix of "y∞" and "u" signals;
$D_{\infty y}$	Transition matrix of "w" and "y∞" signals;
D_{2u}	Transition matrix of "y2" and "u" signals;
D_{2w}	Transition matrix of "y2" and "w" signals;

The above calculations were performed in port, and the route results are shown in Figure 8.8. The study demonstrated the correctness of the synthesis, especially in real conditions where the ship is affected by both wind and wave. Researchers at Gdynia Maritime University are also developing mathematical models of wind and wave as presented in the papers [21,34].

8.3 PSO – Particle Swarm Optimization

The particle swarm algorithm is used for numerical solutions. PSO is included in the group of metaheuristic optimization methods, often also referred to as nature-inspired algorithms or computational intelligence algorithms. As written in [35] and [36], the PSO method is used to solve nonlinear optimization problems with constraints. The objective function for the nonlinear equation is presented below:

$$F(x) = \alpha f_1 + \beta f_2 \tag{8.3}$$

where the functions f_1 and f_2 are the respective values for the calculated standard H_∞ and H_2.

Until now, the coefficients α and β in all calculations had values of 1. However, PSO was introduced to optimize the values of the weighting coefficients in the presented objective function (1). The study was conducted using the Global Optimization Toolbox in the Matlab 2023b environment. The particleswarm command was used during the study: **[optWeights, minLoss] = particleswarm(objective, nvars, lb, ub, options)**

The best weights for the calculated objective function after 20 iterations were found to be 2,47677.

The result of the objective function calculated using the PSO method:

$$F(x) = \alpha f_1 + \beta f_2 = 2.47 \tag{8.4}$$

This means that the numerical values for each coefficient are

$$f_1 = 264.7557, f_2 = 2212.0122 \tag{8.5}$$

$$\alpha = 0.001, \beta = 0.001 \tag{8.6}$$

The α and β values were determined based on personal experience. The state regulator matrix K was synthesized again in the proposed solution, with the parameter content almost five times smaller. Its numerical content is shown below:

$$\mathbf{K} = \begin{bmatrix} 344.2048 & -26.7384 & -6.1341 & -195.3313 & -0.1743 & 0.1038 \\ -22.3865 & 970.3772 & -20.6807 & -0.0724 & -575.0960 & 3.2620 \\ -7.0213 & -21.7927 & 264.5082 & 0.8663 & 0.9445 & -148.5209 \end{bmatrix} \tag{8.7}$$

8.4 Results of automatic ship steering

The ship sailed independently in the port from the quay called ALPHA 1 to the quay ALPHA 2. The results of the simulation carried out are for two studies. The first study is a simulation performed in the MATLAB 2023b environment for the objective function referring to Equations (8.1) and (8.2), where the functions f_1 and f_2 are respectively selected values for the calculated norm $H_\infty = 1.65$ and $H_2 = 242239.8314$ (Figure 8.9)

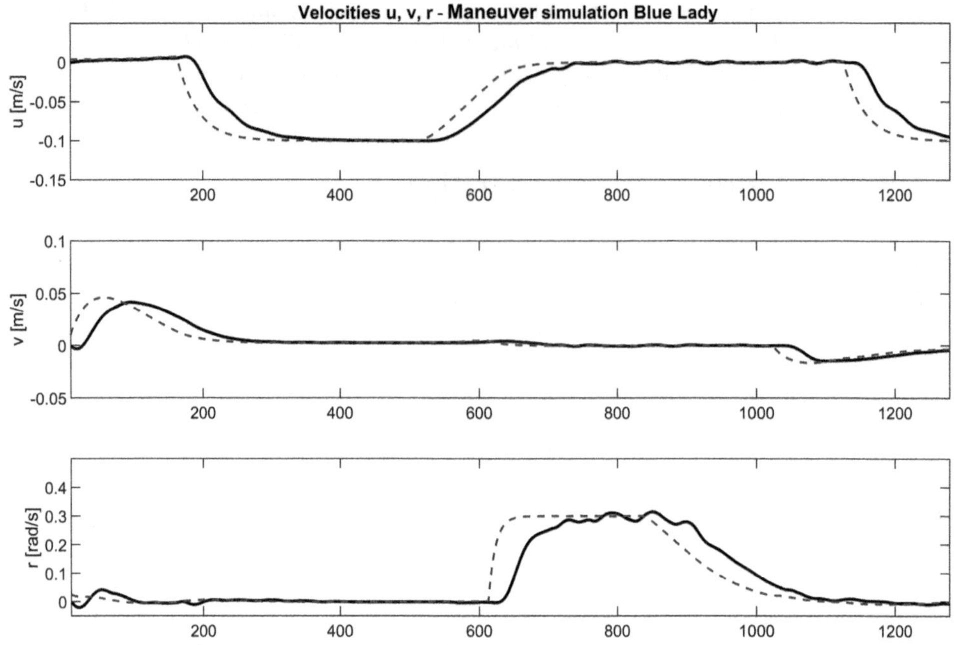

Figure 8.9 Simulated ship maneuver in the MATLAB environment without optimization of the target function. The dashed line set values of longitudinal, transverse, and angular speed. The solid black line represents adjusted values for individual ship speeds.

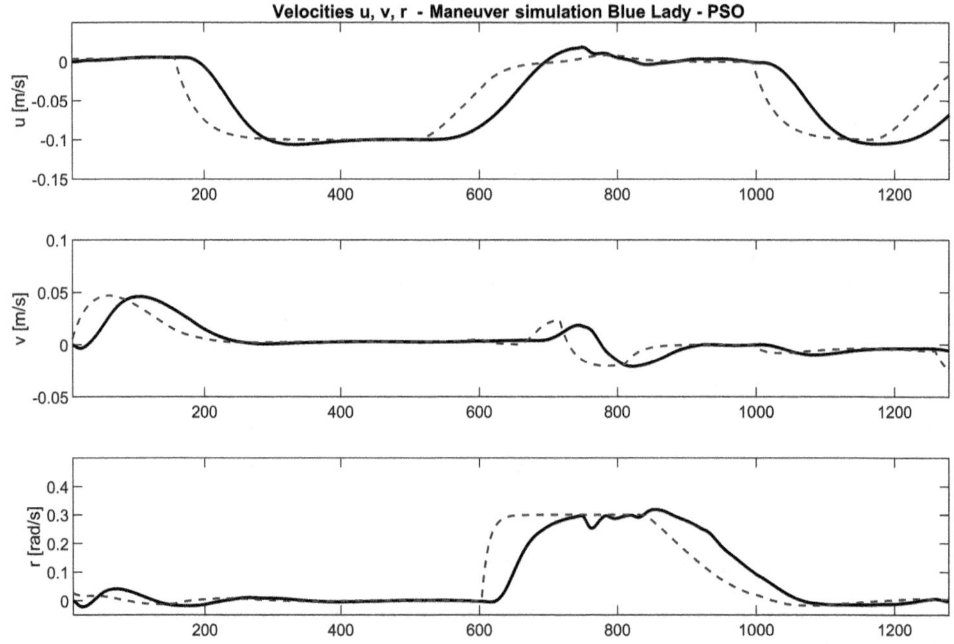

Figure 8.10 shows a simulated ship maneuver in the Matlab environment with optimization of the objective function based on the PSO method. The dashed line represents the set values of longitudinal, transverse, and angular speed, while the solid black lines show the adjusted values for each ship speed.

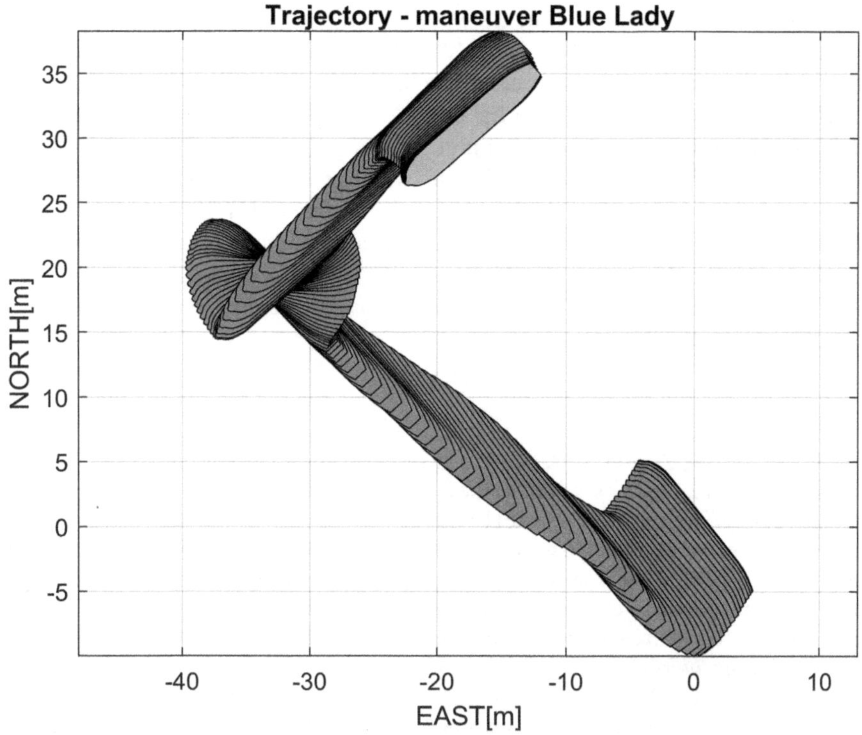

Figure 8.11 The trajectory of the Blue Lady ship was simulated in the MATLAB 2023 environment using the PSO method to find the local minimum for the objective function. The graph displays the optimized trajectory of the ship.

The second study is a simulation performed in the MATLAB 2023b environment for the objective function referring to Equations (8.1) and (8.2), where the functions f_1 and f_2 are calculated based on the PSO method with values for the $H_\infty = 264.7557$ and $H_2 = 2212.0112$ calculated based on the iterations performed (Figure 8.10).

In both charts, red boxes have been introduced to represent the difference in the maneuver performed, and it can be seen in the charts that the longitudinal speed was completed 200 seconds earlier than without minimization. Moreover, the execution of a slight sideways maneuver at the 800th second resulted in a shorter rotation time. Despite identical thruster effort and no change in rotation velocity, the overall rotation maneuver was completed more quickly. This effect is highlighted by the red box on the ship's trajectory chart below (Figure 8.11).

8.5 Conclusion

This study was undertaken to minimize the objective function associated with the control of a training ship's motion in a port environment. Initially, a simulation was conducted without the use of PSO optimization, demonstrating the effectiveness of the synthesis of two controllers (LMI with overriding trajectory control) in guiding the vessel along predefined trajectory points.

Following trajectory analysis, the PSO method was introduced to further minimize the objective function, with particular focus on the H_∞ norm and H_2 norm—representing signal energy optimization and the minimization of regulation error, respectively.

Simulation results confirmed that the PSO evaluates the objective function at each particle's position, identifying both the best (i.e., lowest) value and the optimal particle location. In this experiment, the total time required to complete the target trajectory was reduced by approximately 200 seconds, indicating the potential of the PSO approach for improving performance in similar ship maneuvering scenarios within port areas.

Further testing of the performed simulations with the PSO method should be carried out under real conditions in the port of the Foundation for Safety of Navigation and Environmental Protection in Kamionka, near Ilawa. It is worth considering the application of swarm particle algorithms to other variables, such as calculating the values of the coefficients of the composite variable "s" for placing the poles in the left half-plane of the composite variable "s" for the synthesis of a LMI controller.

References

1. Zhen, R., Gu, Q., Shi, Z., Suo, Y., An improved a-star ship path-planning algorithm considering current, water depth, and traffic separation rules, *Journal of Marine Science and Engineering,* 11(7), 1439, 2023. https://doi.org/10.3390/jmse11071439
2. Miller, A., Ship model identification with genetic algorithm tuning, *Applied Sciences,* 11(12), 5504, 2021.
3. Öztürk, Ü., Akdağ, M., Ayabakan, T., A review of path planning algorithms in maritime autonomous surface ships: navigation safety perspective, *Ocean Engineering,* 251, 111010, 2022.
4. Lazarowska, A., Ship's trajectory planning for collision avoidance at sea based on ant colony optimisation, *The Journal of Navigation,* 68(2), 291–307, 2015.
5. Kennedy, J., Eberhart, R., Particle swarm optimization. In *Proceedings of ICNN'95-international conference on neural networks* (Vol. 4, pp. 1942–1948), Perth, WA, Australia: IEEE, November 27–December 1, 1995.
6. Eberhart, R. C., Shi, Y., Kennedy, J., *Swarm Intelligence*. Elsevier. 2001. https://shop.elsevier.com/books/swarm-intelligence/eberhart/978-1-55860-595-4
7. Lazarowska, A., A comparative analysis of computational intelligence methods for autonomous navigation of smart ships, *Electronics,* 13(7), 1370, 2024. https://doi.org/10.3390/electronics13071370
8. Yu, X., Xiong, X., Wu, Y., A PSO-based approach to optimal capacitor placement with harmonic distortion consideration, *Electric Power Systems Research,* 71(1), 27–33, 2004.
9. Freitas, D., Lopes, L. G., Morgado-Dias, F., Particle swarm optimisation: a historical review up to the current developments. *Entropy*, 22(3), 362, 2020.
10. Song, Y., Cao, X., Review of intelligent ship path planning algorithms, *Frontiers in Management Science,* 3(1), 90–101, 2024.
11. Van den Bergh, F., Engelbrecht, A. P., A study of particle swarm optimization particle trajectories, *Information Sciences,* 176(8), 937–971, 2006.
12. Wei, Y., Qiang, L., A review of particle swarm optimization algorithms, *Chinese Journal of Engineering Science,* 5(3), 87–94, 2004.
13. Kang, Y. T., Chen, W. J., Zhu, D. Q., Wang, J. H., Xie, Q. M., Collision avoidance path planning for ships by particle swarm optimization, *Journal of Marine Science and Technology,* 26(6), 3, 2018.
14. Lyu, H., Hao, Z., Li, J., Li, G., Sun, X., Zhang, G., Zhang L., Ship autonomous collision-avoidance strategies – a comprehensive review, *Journal of Marine Science and Engineering,* 11(4), 830, 2023.
15. IMO org, *IMO,* https://www.imo.org/en/MediaCentre/HotTopics/Pages/Autonomous-shipping.aspx
16. IMO, https://www.imo.org/en/MediaCentre/PressBriefings/Pages/08-MSC-99-mass-scoping.aspx
17. Wang, L., Wu, Q., Liu, J., Li, S., Negenborn, R., State-of-the-art research on motion control of maritime autonomous surface ships, *Journal of Marine Science and Technology,* 7, 438, 2019.
18. Fundacja Bezpieczeństwa Żeglugi i Ochrony Środowiska Web site FBŻiOŚ, https://www.ilawashiphandling.com.pl/

19. Google maps, https://earth.google.com/web/data=MkEKPwo9CiExMTZBQWJWR2Zvc2hNRW1DdFdYYzZ0cEFLQ0ExM0h4d1oSFgoUMDBFMTVCRUE1NTMyRjE1QzFGNjYgAQ
20. infoIlawa zdjecie nabrzeza z artykulu: https://www.infoilawa.pl/aktualnosci/item/75753-pod-ilawa-dziala-najlepsza-na-swiecie-szkola-kapitanow-i-pilotow-zeglugi-wielkiej-zdjecia-wideo#&gid=1&pid=1
21. Gierusz W., Rybczak M., Effectiveness of multidimensional controllers designated to steering of the motions of ship at low speed, *Sensors,* 20, 3533, 2020.
22. Gierusz W., Rybczak M., Maritime autonomous surface ships in use with LMI and overriding trajectory controller, *Applied Sciences,* 12(19), 9927, 2022.
23. Gierusz W., Miller A., Ship motion control system for replenishment operation, *Applied Mechanics and Materials,* 817, 1197–1214, 2016.
24. Koznowski, W., Kula, K., Lazarowska, A., Lisowski, J., Miller, A., Rak, A., Rybczak, M., Mostefa, M.-S., Tomera, M., Research on synthesis of multi-layer intelligent system for optimal and safe control of marine autonomous object, *Electronics,* 12(15), 3299, 2023.
25. Miller, A., Rak, A., Measurement system for the environmental load assessment of the scale ship model, *Sensors,* 23(1), 306, 2022.
26. Rybczak, M., Rak, A., Prototyping and simulation environment of ship motion control system, *TransNav,* 14, 367–374, 2020.
27. Tomera, M., A multivariable low speed controller for a ship autopilot with experimental results, in *2015 20th International Conference on Methods and Models in Automation and Robotics (MMAR), Międzyzdroje* (pp. 17–22). *IEEE,* 2015.
28. Tomera, M., Hybrid switching controller design for the maneuvering and transit of a training ship, *International Journal of Applied Mathematics and Computer Science,* 27(1), 63–77, 2017.
29. Lazarowska, A., A comparative analysis of computational intelligence methods for autonomous navigation of smart ships, *Electronics,* 13(7), 1370, 2024. https://doi.org/10.3390/electronics13071370
30. Rybczak, M., Cieślikiewicz, W., Effect of lake waves on manoeuvring performance of a model ship. *Ships and Offshore Structures*, 19(10), 1726–1739, 2024.
31. Rybczak, M., Lazarowska, A., Safe control of an autonomous ship in a port based on ant colony optimization and linear matrix inequalities, *Electronics,* 13(17), 3525, 2024.
32. Rybczak, M., Lazarowska, A., The development of algorithms for safe control of an autonomous ship, *International Journal of Electronics and Telecommunications,* 70, 997–1003, 2024.
33. Gierusz W., *Synteza Wielowymiarowych Ukladow Sterowania Precyzyjnego Ruchem Statku z Wykorzystaniem Wybranych Metod Projektowania Ukladow Odpornych*, Gdynia: Akademia Morska w Gdyni, 2004.
34. Rak, A., Miller, A., Modelling of lake waves to simulate environmental disturbance to a scale ship model, *Polish Maritime Research,* 3, 12–21, 2023.
35. Foryś P., Zastosowanie metody roju cząstek w optymalnym projektowaniu elementów konstrukcji, *Czasopismo Techniczne. Mechanika,* 105, 31–42, 2008.
36. Marini, F., Walczak, B., Particle swarm optimization (PSO), a tutorial. *Chemometrics and Intelligent Laboratory Systems*, 149, 153–165, 2015.

Chapter 9

Information flow in transport telematics systems

Mirosław Siergiejczyk

9.1 Introduction

The main objective of the transport system is to move people and/or cargo, which is determined by the type, number, and characteristics of the objects being moved, as well as transport relations and transport service quality parameters (safety, speed, comfort, etc.). This objective is achieved by using means of transportation in the transportation process. For the transportation process to proceed with specified efficiency and required safety, the transport system is supported by many subsystems, including telematics systems. The input to the transport system usually consists of a stream of technical and organizational resources necessary for its functioning such as materials, energy, technical equipment (devices and machines), personnel, and financial resources. Most importantly, it includes transport needs identified by the environment. The output of the transport system is the desired spatial location of the objects being moved, the characteristic of which is the size of the transport needs fulfilled, together with an assessment of the quality of their implementation. The environment of the transport system is composed of sectors of the national economy other than transport, society, and the natural environment. The transport system model is a representation of inputs and their transformation so that the outputs, according to the adopted criteria, are assessed as completed tasks (services) with an appropriate level of their quality [1].

Transport telematics is a branch of knowledge that integrates computer science and telecommunications in applications for the organization, management, and control of traffic in transport systems. It stimulates technical and organizational activities that ensure the quality of transport services, improving the efficiency and safety of these systems [2].

The key functions of telematics systems are information management functions. This involves the acquisition, processing, distribution together with transmission, and use of information in decision-making processes. Such processes are either carried out in a predetermined manner (e.g., automatic traffic control) or result from ad hoc situations (decisions of dispatchers, independent users of infrastructure requiring support with current information) [3].

The basic feature of telematics applications is the ability to deliver and process a rich set of information appropriate for a given application, adapted to the needs of recipients – users of this information, at the right places and the right time. Information operations can be carried out either automatically or interactively at the specific request of the user. An important feature of telematics applications is the ability to effectively associate the operation of various subsystems and integrate them into a coordinated mode of operation. The aim of this work is to present transport telematics systems in the aspect of transferring and transmitting information, taking into account the specific requirements resulting from supporting transport processes.

DOI: 10.1201/9781003686088-9

9.2 Transmission of telematics information

One of the most essential properties of telematics systems is the transmission and flow of information. Telematics information flow is directly related to telecommunications, which involves the transmission of messages over a distance using various types of signals, most commonly electrical or optical signals nowadays. Messages transmitted via telecommunication services can be in multiple forms. We can distinguish alphanumeric data, voice, sounds, moving and non-moving images, written characters, and various types of measurement signals. When considering the flow of telematic information in the telecommunications sense, it is necessary to pay attention to the information chain. The essence of the information chain is the transmission of various types of messages from the sender to the receiver, thus focusing on these two points of data exchange. The manner of the transmission itself is not important, only its occurrence. The manner of transmission of information is important in the case of the information transmission chain, which is contained in the information chain (Figure 9.1). In the case of an information transmission chain, what is important is the transmission of data from the sender to the receiver without identification. In this case, what matters is the transformation of the message into a transmittable form and the connection path between the points in question. During transmission, the signal is subject to various types of interference, which often causes distortion. Therefore, it is important that the signal delivered to the recipient is the best representation of the original signal. The connection path between two network devices is referred to as the teletransmission path, which is most often in the form of wire (symmetrical, coaxial), fiber optic, or radio [4].

The telematics system interacts with many other systems and the external environment, and may include many component subsystems. This results in the need for information exchange between the components mentioned (Figure 9.2) [4]. To ensure the realization of data flow, it is indispensable to use various types of transmission media, without which the transmission of information would not be possible. Transmission media are used to ensure fast and reliable communication with extensive systems that require the transmission of large amounts of data over long distances, as well as for the transmission of simple control messages or measurement data from sensors over short distances. Therefore, transmission media and information transfer mechanisms play a significant role in the systems in question.

The central point of the entire telematics system is the dispatch center (management center). A block diagram illustrating the flow of telematics information is shown in Figure 9.3. The specificity of information transmitters and receivers requires the use of different methods and media for data transmission. The selection of the method and the appropriate transmission medium depends on the devices used and their technical advancement, requirements for throughput, power supply, etc. In addition, weather conditions, the possibility of electromagnetic interference,

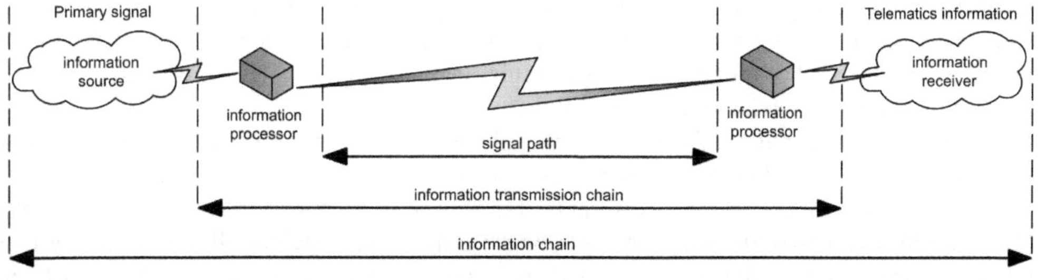

Figure 9.1 Telematics information chain and telematics information transmission chain [4].

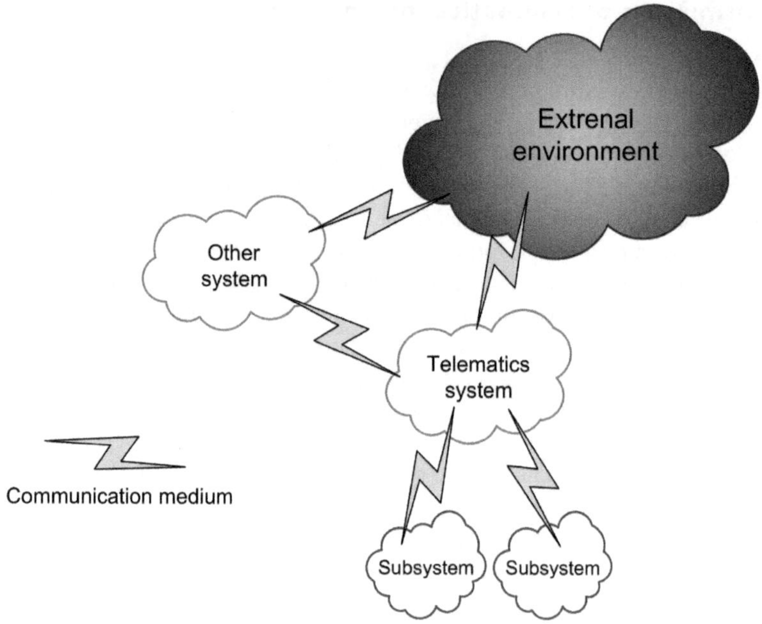

Figure 9.2 Telematics information exchange.

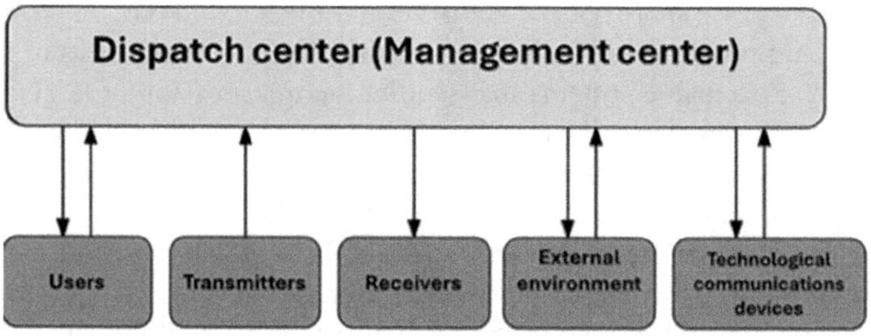

Figure 9.3 Flow chart showing the flow of telematics information.

mobility or stationarity of elements, working conditions, and requirements resulting from the software and telecommunications infrastructure are also important. One of the most important factors is also the cost of implementation and subsequent operation of the selected medium and method of transmitting telematics information.

An example of telematic information flow is the exchange of data between highway section management centers and highway telematics systems. The highway section management centers are the main and most important elements of the highway telematics infrastructure (Highway Management Center/Traffic Management Center). The center receives all data from the telematics systems used on the motorway and manages and operates these systems from there. Variable message signs and cameras along the monitored section of the road can be controlled using special hardware and software. The center houses an emergency call center and a system for collecting and analyzing weather data. It also performs functions related to comprehensive motorway monitoring and sends appropriate messages to users (drivers) [4,5].

Among the tasks performed by the motorway section management center, we can distinguish:

- vehicle traffic management,
- supervision of weather and traffic conditions on a given section,
- control of road surface parameters,
- management of emergency situations (accidents) and handling of emergency calls,
- provision of information for travelers both before and during the journey,
- visual supervision of the motorway and response in the event of threats,
- management of maintenance services.

An important aspect of the motorway section management center is its location in the field. Centers are located approximately every 60 km, ensuring that the distance from the site of an adverse incident to the nearest center is at most 30 km. This is also related to the fact that maintenance and emergency service facilities are often located nearby. However, in some cases, large motorway surveillance and traffic management centers are established to cover several hundred kilometers of road. It is also important for the center to be located in or near large population centers for easy sourcing of skilled staff and availability of infrastructure elements.

To ensure the efficient operation of motorway section management centers, it is necessary to ensure the transmission of information. This is achieved through communication with motorway telematics systems (alarm communication, weather coverage, and video surveillance) and the exchange of information with other management centers, primarily facilitated by optical fiber running along the motorway. The signals are then appropriately converted, decoded, and read by specific applications.

Communication with emergency services is also important and must be direct. The connection is established immediately after the operator is called. A telephone line with a "hotline" service is most often used for this purpose. Radio communication, such as CB – Citizen-Band with a wide range, is used for communication with maintenance units located in the field.

9.3 Communication architecture of transport telematics systems

9.3.1 Determination and tasks of the communication architecture of telematics systems

The communication architecture of transport telematics systems defines and determines the structure and means of enabling the exchange of information between different parts (elements) of the system (means of transmitting data streams). This concerns two complementary aspects covered by separate procedures:

- providing the means to enable point-to-point data transfer,
- ensuring the reliability of transmission and interpretation of telematics information.

The problem of providing the means to transfer data from point to point and ensuring that the selection of these means is appropriate in terms of cost and acceptable distortion or delay of the transmitted data concerns the description and definition of the means of transmitting information. In telecommunications terms, this pertains to the first four layers of the OSI (*Open Systems Interconnection*) model: the physical layer, the link layer, the network layer and the transport layer [1]. The communication platform must remain as independent as possible from the technologies and specific products used. To fulfill this requirement, solutions such as analyzing physical data stream flows in representative sample systems and determining the most essential telecommunications needs in the telematics system based on this analysis, considering the necessary interfaces, are envisioned. Description of typical telecommunications needs is an important advantage of the communication architecture. Telecommunications technologies change very quickly and it is not possible to develop a technology-independent architecture that is up to date for a long period of time. However, the proposed general description of a typical data flow will remain valid as long as the image of the most representative sample systems does not change significantly. Therefore, this methodology provides a good basis for the analysis of issues relating to telecommunications in ITS systems in general [5].

In addition to purely telecommunications issues, the problem of ensuring the reliability of telematics information transmission and its correct interpretation should also be taken into account. This is the problem of standards that would provide the communicating parties with the ability to exchange data in a maximally reliable and efficient manner. However, in reality, the ability of the system to provide services is largely dependent on the data that the system can process in a unit of time. Therefore, it is important to consider not only the type of data processed in the system but also the level of detail available for use in the construction of a specific service or other operation needed in the telematics system. Other technical problems, such as the location of data storage and processing, cannot be omitted either. This solution may lead to the need to diversify the methods of exchange, and thus,– for example,– the use of different standards for presenting messages.

In the communication part, telematics systems define the connections between the transport environment and the telecommunications environment. Due to the dynamic development of telecommunications, telematics system designers have a wide choice of telecommunications means, which allows them to consider the needs resulting from various implementation circumstances. However, the communication architecture does not provide guidance on specific systems or technologies, but only identifies the possibilities created by the technologies used. Four types of media have been distinguished here, which can be used to transmit information in transport telematics system solutions. These are telecommunications systems [6]:

- wired (fixed point – fixed point communication),
- long-range wireless (stationary point – mobile point communication),
- dedicated short-range (stationary point – mobile point communication),
- between mobile terminals in means of transportation (mobile communications).

It should be noted that there are multiple transmission techniques used for communication between fixed points. For example, proprietary or leased pairs (twisted pairs), coaxial cables, and optical fibers may be used to support the traffic control subsystem. In other applications, microwave lines, spread spectrum radio systems, or local radio networks may be more practical. Despite their use to support fixed points, they are treated as wireless media in the architecture.

In the physical architecture structure, it is envisioned that the subsystems within the Dispatch Centre (Management and Supervision) are connected to each other by a wired network. This enables each subsystem to collect, integrate, and distribute information to the other subsystems with mutually recognized communication and coordination rules, enhancing the efficiency of the subsystems' operations. The architecture distinguishes between two types of wireless communication depending on coverage and territorial scope. The first is long-range wireless communication, which defines the means used for services and applications when information is distributed to users distant from the source of this information, requiring continuous network availability coverage in a defined area. A further distinction is made between one-way and two-way communication, as this affects the choice of technology (e.g., in one-way there may be radio). The second type is short-range wireless communication, used to transmit information of a local nature. There are two types here: vehicle-to-vehicle (V2V) and DSRC (dedicated short-range communications). The first type is used for collision avoidance applications, while DSRC is used for electronic toll collection, access authorization, etc. In general, it can be concluded that an analysis of the telecommunication traffic loads generated and the time requirements for the transmission of information, as well as an analysis of the available transmission techniques, indicate that existing commercially available wired and wireless networks are capable of meeting current and future needs for the transmission of telematic information.

9.3.2 Communication architecture in road transport

The communication architecture framework presented in the European project KAREN (Keystone Architecture Required for European Networks) describes mechanisms supporting the exchange of information between different parts of the system. It is assumed that this exchange should meet two conditions:

- The mechanism should enable data transmission between interested points of the system and be balanced in terms of cost, accuracy, and transmission delays.
- It should ensure the correctness of information transmission and interpretation.

Both of these conditions require:

- introducing a detailed analysis of the definition and description of communication connections in the main interfaces of the transport physical structure subsystems,
- defining the necessary communication protocols.

The physical structure of the system, that is, the identification of elements that exchange information, is categorized in KAREN as Physical Architecture. Five main types of elements are identified here, and the requirements for information communication between such elements are characterized [6,7] as follows:

- **Center**: a place where traffic data, charges, and transport orders are collected and comparatively validated, and where this data is stored. It is also where initiations for traffic measurement and instructions for rolling stock management are generated (e.g., traffic control centers, traffic information centers, rolling stock management centers);
- **Kiosk**: an installation usually located in a public place, offering information facilities for travelers (e.g., a tourist information point, often self-service);
- **Roadside**: a place where traffic, vehicles, and pedestrians converge, where charges may be collected, measurements for traffic management are taken, or information for travelers is provided; may also serve as a tourist information point, often self-service);
- **Roadside**: a place where traffic, vehicles, and pedestrians appear, tolls are collected, measurements necessary for traffic management are taken, or information for drivers and pedestrians is provided;
- **Vehicle**: a device that can move on the road network and transport one or more persons (e.g., bicycle, motorbike, car, public transport vehicle) or goods (any type of freight vehicle);
- **Traveler**: a person driving a vehicle or using a vehicle.

Since connections to the environment require a systemic view of the description of that environment, its active elements are defined as terminators. A terminator can be a person, system, or entity from which data can be received and to which data requests are sent accordingly. These may include, in particular:

- road infrastructure (roadside),
- emergency systems,
- external service providers,
- meteorological systems,
- multimodal transport structures,
- travelers.

Figure 9.4 shows an example of information exchange in an integrated traffic and road transport management system.

Figure 9.4 Information exchange in the integrated traffic and road transport management system. (Based on [6].)

The description of information communication requires distinguishing:

- the communication link, that is, the means of transmitting physical data
- the location, source, and outlet of data, that is, the places where the data originate or to which they are directed,
- the physical interfaces, that is, the parts of the system through which the information exchange is realized.

A systematic characterization of information exchange processes requires dividing this area into two parts:

- Internal communication within the system, including the exchange of information between its variously located elements;
- Communication between the system and external entities.

In vehicle-to-everything (V2X) communication, all vehicles and infrastructure are interconnected. The concept of V2X covers many possible applications, which can be categorized. These include V2V, Vehicle-to-Infrastructure (V2I), Vehicle-to-Network (V2N), and Vehicle-to-Pedestrian (V2P) communication [4].

The use of V2X systems is primarily about increasing traffic safety and efficiency. Looking at the practical side, the technology will enable vehicles to, among other things, communicate with other vehicles and with the road infrastructure. Vehicles will send and receive information about heavy traffic in the area or random events such as accidents and other traffic obstructions. The concept also includes access to the Internet, which will enable the use of a web browser, watching online videos, or downloading files and applications. Cars equipped with V2X will also be subject to continuous diagnostics and be able to be remotely managed and controlled – important, for example, in situations where we have to travel to an area that is dangerous to human life [8].

There are currently two main access methods for V2X communication. The first is based on wireless virtual LANs (WLANs) – DSRC technology and the 802.11p standard, while the second is cellular networks [9,10].

Work on the implementation of V2X has been ongoing for many years, and its implementation is built on the LTE interface, which is sufficient for many services and meets the requirements for a large part of the defined possible applications of the technology. However, the use of 5G NR is essential for the implementation of services that require very low latency, reliability, longer transmission ranges, or high data rates. In favor of mobile networks – particularly the fifth generation – is their functionality, and more specifically their layer partitioning (Network Slicing). This involves the virtual division of the network in such a way that, although various services will be supported by the same backbone network, the individual services will be separated from each other – grouped into layers (Slices). Services in each layer will operate and transmit their data independently, and the division of services into layers will depend on what resources they require – for example, the degree of privacy, acceptable latency, transmission rate, or frequency bandwidth [9].

Cellular networks, particularly those based on the 5G NR system, are a solution that will enable the deployment of smart services in the transport sector and, in the future, will support the deployment of fully autonomous vehicles based on, among other things, V2X communication [8]. The high demands placed on the next generation of networks are key to the development of intelligent transport [10].

Dedicated short range communications (DSRC) is a system that includes a number of standards, mainly developed by the IEEE organization. DSRC was developed for the needs of the transport industry. It is based on the IEEE 802.11 standards that define Wi-Fi networks. DSRC is therefore a wireless communication and information exchange system for automotive vehicles. The aim of developing this system is to connect all vehicles to each other and to the surrounding infrastructure – V2V and V2I communication [11,12].

When DSRC is mentioned, the Wireless Communication for Vehicular Environment (WAVE) standard should also be mentioned. These terms are often used interchangeably, but DSRC is a slightly broader concept. WAVE is a group of standards from the IEEE 1609.x series and the IEEE 802.11p standard (Figure 9.5).

An important consideration when designing a communication system for the transport industry was to create an efficient, yet low-cost system. The developers, therefore, based their design on the Wi-Fi standard, which uses unlicensed frequency bands, that is, those that anyone can use without any charge. The 5.9 GHz band has been reserved for the DSRC system in Europe [11].

The need to implement an intelligent communication system in the transport environment is due to the increasing volume of traffic and the rise in accidents. This leads to higher human fatalities and significantly increased material costs. The growing traffic volume also has a negative impact on the climate. Poorly organized traffic causes congestion and gridlock on the roads, leading to higher emissions and unnecessary waste of fuel resources. The implementation of an intelligent system will have a positive impact on many aspects, including reduced fuel and electricity consumption, better traffic organization and management, good communication between road users, less congestion, and much greater road safety. The system will achieve this with the help of short messages sent between vehicles and between the vehicle and the infrastructure.

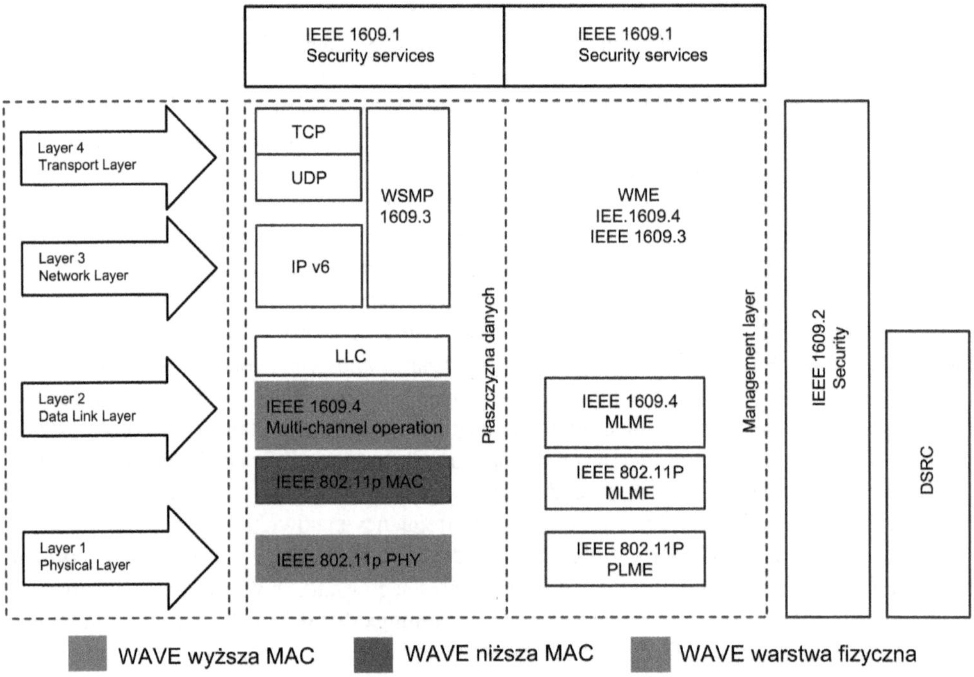

Figure 9.5 WAVE standard protocols. (Based on [13].)

9.3.3 Comparative analysis of communication technologies DSRC vs. C-V2)

Both DSRC and C-V2X are undergoing extensive improvements to support advanced in-vehicle applications with high reliability, low latency, and high bandwidth requirements. These evolutions of RAT (radio access technology) – IEEE 802.11bd for DSRC and NR (New Radio) and V2X for C-V2X – can complement today's in-car sensors to enable autonomous driving [14].

Another important issue in road transport telematics systems is the identification of communication threats in the V2V and V2I information exchange models, considering the already known threats affecting sensory systems, especially vehicles. The analysis of selected information exchange threats in these communication models was conducted, among others [15,16].

A comparative analysis of DSRC and Cellular Vehicle-to-Everything (C-V2X) requires a multi-dimensional evaluation of both technologies across several key aspects such as communication range, latency, reliability, scalability, interoperability, spectrum usage, deployment status, and long-term viability. These technologies are central to V2X communication paradigms in Intelligent Transportation Systems (ITS), enabling interactions among vehicles, infrastructure, pedestrians, and the network.

9.3.3.1 Overview of DSRC and C-V2X

DSRC

- Based on the IEEE 802.11p standard, an extension of Wi-Fi (IEEE 802.11) tailored for vehicular environments.
- Operates in the 5.9 GHz band, with a range of up to 300–1,000 meters.
- Enable direct communication without the need for cellular infrastructure.
- Initially championed in the United States and supported by the USDOT for V2V and V2I applications.

C-V2X (Cellular Vehicle-to-Everything)

- Standardized by 3GPP in Release 14 (LTE-V2X) and later extended in Release 16 and beyond (5G NR-V2X).
- Supports two modes:

  PC5 (sidelink) for direct communications (like DSRC),
  Uu interface for network-based communications via cellular base stations.

- Operates in the 5.9 GHz band (like DSRC) or in licensed cellular bands.
- Backed by major telecom and automotive industry players (Table 9.1).

DSRC is a mature, dedicated V2X communication technology with proven safety application support but limited scalability, spectral efficiency, and adaptability to future networks. C-V2X, particularly in its 5G NR-V2X evolution, presents a more flexible, scalable, and integrated solution that aligns with the broader 5G ecosystem, Mobile Edge Computing (MEC), and smart city architectures. While DSRC may persist in niche deployments or regions with existing infrastructure, the industry consensus is clearly shifting toward C-V2X as the long-term V2X communication standard. The future of V2X communication appears to be dominated by C-V2X,

Table 9.1 Comparative dimensions

Dimension	*DSRC (IEEE 802.11p)*	*C-V2X (3GPP Rel. 14–16)*
Communication Model	Peer-to-peer, broadcast (without infrastructure)	Hybrid: direct (PC5) and network-based (Uu via eNodeB/gNodeB)
Spectrum	5.850–5.925 GHz (USA), 5.875–5.905 GHz (EU)	Same spectrum for PC5; Uu uses licensed LTE/5G bands
Latency	~1–10 ms	Lower in 5G NR-V2X (sub-1 ms achievable), LTE-V2X: 10–20 ms
Range	~300–1,000 m	Comparable or higher (~500–1,500 m for sidelink, longer via Uu)
Reliability	Medium (contention-based MAC: CSMA/CA)	Higher (semi-persistent scheduling in LTE-V2X)
Network Dependency	Independent of infrastructure	Direct mode: independent; Network mode: requires cellular infra
Scalability	Limited in high-density environments	Better due to scheduled access, particularly in 5G NR-V2X
Interference Handling	Basic mechanisms	Advanced interference and congestion control
Security Framework	Mature via IEEE 1609.2	Under development within 3GPP and ETSI
Ecosystem Support	Limited; largely static since 2010	Growing rapidly; broad industry and 5G integration support
Deployment Status	Pilots in the USA, Japan; some regional use	Global trials, wide support in China, EU, and US (FCC reallocation)
Future Readiness (5G)	Not evolving toward 5G	Fully aligned with 5G roadmap and MEC integration

primarily due to its inherent ability to evolve with 5G technology [14]. The capabilities of 5G NR-V2X, such as enhanced mobile broadband, ultra-reliable low-latency communication, and massive machine-type communication, are seen as essential for realizing the full potential of connected and autonomous vehicles. While DSRC may continue to exist in some legacy systems or niche applications, C-V2X is positioned to be the primary enabler of future ITS.

9.3.4 *Rail transport telematics*

The dynamics of transport and cooperation with other institutions and enterprises mean that transport cannot lag behind the trends of global development in the field of using teleinformatic solutions. If only for reasons of the environment and cooperation with this environment, transport cannot be an "open-air museum" of telematics systems. A common feature of telematics systems, especially those created for transport purposes, is that their operation covers large areas, and therefore they must use data transmission networks.

Railway transport uses telematics for the following needs [17]:

9.3.4.1 Management

Railway companies cannot lag behind their customers' systems in their management systems, and this will require the use of management methods supported by telematics adapted to the specific profiles and requirements of individual railway companies, such as the infrastructure operator, carriers, regional transport, and other companies.

9.3.4.2 *Steering*

Management includes all types of dispatching, such as traffic dispatching, energy network dispatching, and supervision of telecommunications network systems, such as the GSM-R system.

9.3.4.3 *Monitoring of railway areas*

Telematic systems are used to monitor various railway areas, such as marshaling yards, transshipment points, border crossings, railway infrastructure, rolling stock, and means of transport. The telematic services requires by railway customers include monitoring selected railway areas (e.g., passenger parking lots), ensuring passenger safety before, during, and after the journey, providing broadband Internet access, dynamic timetables, mobile information, ticket sales, and monitoring freight transport.

To meet all the requirements of telematic systems implemented for the needs of the railway, an ICT network is needed, which will be the basis for all telematic systems used in the railway.

Currently, construction of a fiber optic network is underway for the needs of the GSM-R system, which is a component of the ERTMS system. The GSM-R network will also replace the technically outdated analog simplex VHF radio communication network operating in the 150 MHz band used so far. The GSM-R network will support approximately 18,000 km of the railway network. In addition to voice transmission services, the GSM-R network also provides data transmission services, which are necessary for modern rail traffic control systems (ERTMS). The wired part of the GSM-R system, connecting its stationary elements (base stations – BTS, base station controllers – BSC, exchanges – MSC), should, for safety reasons, be made on separate fiber optic fibers. On these fibers, appropriate teletransmission systems such as SDH systems and GbEthernet will be implemented [17,18]. Other telematics systems necessary for the operations of railway companies should use different fiber optic fibers located in the same cable as the fibers intended for the GSM-R system.

Progress in wireless networks is also happening rapidly. Currently, a roadmap has been developed defining the time horizon for the implementation of the 5G system for public use networks. Pilot implementations of the 5G system are being carried out by telecommunications companies in various countries, including Poland.

The UIC FRMCS railway mobile communication system, FRMCS, is an acronym for Future Railway Mobile Communication System and is a comprehensive mobile communication system developed by the International Union of Railways (UIC). It is aimed at revolutionizing the way railway vehicles communicate with each other and with dispatch centers [19].

FRMCS is an advanced mobile communication system that provides not only voice communication but also data transmission, vehicle location, traffic management, and travel safety. It is a system designed specifically for the railway sector, integrating various communication and operational functions into a single platform. The main assumption of FRMCS is to create a modern, uniform communication system that will replace existing solutions, such as GSM-R (Global System for Mobile Communications – Railway) and analog radio communication. This system is intended to be open, flexible, and scalable, enabling the integration of various technologies and ensuring reliable communication in various conditions and situations. FRMCS aims to enable uniform communication between different railway systems in various countries, which will enhance interoperability and cooperation at the international level [20]. The transition to a modern communication system will facilitate better monitoring of rail traffic, quicker response to emergency situations, and improved coordination of activities, ultimately leading to increased travel safety. FRMCS is to ensure more effective management of rail traffic, enabling the optimization of schedules, routes, and the use of rail infrastructure. The FRMCS system is a platform that allows for the introduction of new technologies and innovative solutions in the rail sector, such as automation, artificial intelligence, and data analysis. The implementation of FRMCS is intended to contribute to the modernization of the rail sector, enhancing its competitiveness and improving the travel experience for passengers.

9.3.5 Communication architecture for rail freight transport

Providing transport services at an appropriate level in the field of freight transport is a major problem that requires interfaces for the exchange of data (information) between participants in the transport chain. This also applies to railway companies, where information is exchanged between various managers of the railway infrastructure and transport companies. Traffic conducted across state borders requires not only the exchange of data (information) regarding the transported goods but also technical data regarding the train and wagons.

In simplified regulations, the exchange of data related to traffic management between infrastructure managers, the exchange of data regarding the carrier of goods, that is, the organization of goods transport with the appropriate waybill (customs, agricultural goods, and veterinary certificates), and appropriate authorization is required. Specific information from border crossings is sent to transport managers (freight transport management centers, carriers, and logistics service operators) and is used for the effective management of transport processes. Transport managers also exchange information with selected clients who need data for the precise planning of freight transport operations, including loading and departure times.

Each railway company, client, and operator has information about the location of the wagon and the condition of the load immediately after crossing the border. A system in which information (data) is quickly exchanged between partners increases the efficiency of freight transport [7].

An important issue is finding a solution for data transmission in the area of operator networks. Public networks can be used for this purpose, but they are not always adapted to the specific requirements of transport. This mainly concerns ensuring the continuity of service provision (so-called service availability) and ensuring the required level of quality of data transmission services (QoS Quality of Service). It is also possible to utilize the existing and developing infrastructure of telecommunications operators that provide services for the needs of rail technological communication. The international CESAR system developed by the International Union of Combined Road-Rail Transport Companies (UIRR), utilizes the public Internet network to exchange information between road and rail transport operators. Data is transmitted via a network with appropriate security measures using cryptography and suitable access technologies. However, the security of the data concerning train composition, wagon and goods characteristics has not been fully tested yet.

The International Union of Railways (UIC) has developed recommendations in the 407 series [21] regarding the exchange of data in traffic crossing national borders. However, these recommendations are not mandatory and, as a result, are implemented to a small extent by national freight carriers. The implementation of a properly designed teleinformatics platform can contribute to increasing the scope of transmitted data as well as the speed of information exchange and can have a significant impact on increasing the efficiency of transport, particularly multimodal transport.

Designing an appropriate teleinformatics platform requires solutions in the format of transmitted data, the technique of their transmission, and the selection of the transmission network technology. Figure 9.6 shows the general structure of the teleinformatics platform for the needs of storing, commuting, and transmitting information related to the implementation of cargo and goods transport processes [7].

All participants in the transport chain are connected to each other through a central platform (central in the logical sense; physically the "center" may be dispersed). Connections to the central platform are primarily based on the existing capabilities of individual users. The platform does not process information that will be exchanged in the user standard. If necessary, this standard will be provided from the outside. Therefore, for the platform to work correctly, it is not necessary for all users to use the same standard, which will affect the speed of implementing the platform into operation. Transport participants provide information to the platform, and the platform sends it to specific users connected to it. It is worth noting that technical conversions between the standards of platform subscribers are carried out automatically through the platform, without interfering with the data content being exchanged. IT technologies such as electronic signature and coding should comply with the requirements, ensuring the confidentiality of information and authentication of electronic documents along with the electronic signature. For the purpose of implementing data transmission services in the proposed "structure of electronic freight transport" in Poland, and for the needs of freight transport crossing borders, justified

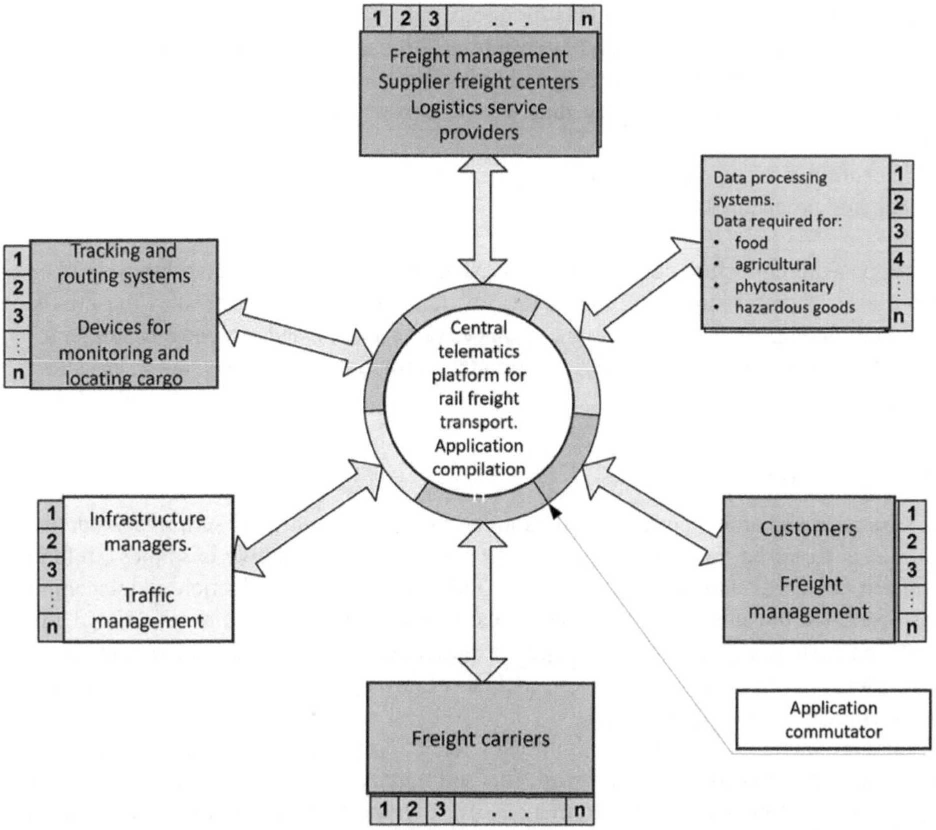

Figure 9.6 General structure of communication architecture in the freight transport chain [7].

by services and technically, the use of the operator network of Telekomunikacja Kolejowa is necessary. Telekomunikacja Kolejowa Sp. z o.o. is a public telecommunications operator and supplier of comprehensive communication and data transmission services used to manage, control, and monitor train traffic in companies in Poland [7].

Descriptions of data required in rail freight traffic are specified in the Technical Specification for Interoperability – Telematic Applications for Freight Transport (TSI-TAF), while for passenger transport in the Technical Specification for Interoperability – Telematic Applications for Passenger Transport (TSI-TAP) [21,22].

In addition to providing information for modern logistics, railway telecommunications networks must transfer existing information related to control and management of rail traffic, as well as the management of railway companies. The role of railway telecommunication networks is evolving, and their development must keep pace with the advancements in modern logistics services and forms of enterprise management.

Electronic information exchange guarantees rapid exchange of information, authentication, and security, providing the opportunity for all participants in goods transport to be integrated into one organization. The advantages of such a solution are:

- Reduction of transit time,
- Improvement of wagon turnover,
- Reduction of the number of wagons for goods transit,
- Improvement of communication between transport users through electronic documents,
- Transparency of transport processes and better planning of transport tasks,
- Transport between large railway junctions is unrestricted by technical and organizational conditions,
- Concentration of management and service in freight traffic centers, where data is accumulated, and procedures are launched.

The proposed architecture concept for data exchange can enhance transparency in operations for all tasks, thereby improving the quality of transport services. It enables the provision of continuous and up-to-date information, enhances rail services, and can boost customer satisfaction with the transport services offered, ultimately attracting them to use rail services for future transport needs.

9.4 Summary

The structures and architectures presented in the study are primarily descriptive in nature. Their proposal is based on the analysis of existing systems (representative examples), reference to European standards and projects (KAREN, TSI, UIC), logical deduction of telecommunications needs and the author's proposal (ICT platform for railways). The methods used are based mainly on system engineering and expert experience, and not on methodologies for formal verification of system architecture (e.g., analysis of throughput, reliability, resistance to attacks, energy efficiency, QoS indicators).

Transport telematics systems are built using technologies at various stages of development. This results in the need to conduct work focused mainly on the problems of acceptance, system integration (integration of information, overcoming technical inhomogeneities, creating uniform system platforms), functioning in real conditions, and methods of their operation. An important issue is the assessment of the use of telematics systems in the implementation of tasks in the transport process, considering the operational properties of these systems. The efficiency of transport telematics systems is significantly influenced by their technical condition, readiness to perform information transmission tasks, external conditions, and the broader environment related to the transportation of passengers and goods.

The expanding use of telematics systems creates a potential risk in the proper functioning of transport telematics systems in the future, which is still difficult to estimate. The "networking" and integration of computer systems, driven by the introduction of telematics, pose an increasing threat of new types of attacks, such as exploitation of network access or deliberate destruction of system components. In the further development of telematics, the "Fail-Safe" principle should also be applied in the future.

Telematics systems require expensive technical components such as measuring and information equipment, telecommunications devices, software. These costs burden both system operators and their users. Cost reduction would be possible by integrating telematics functions and using them more fully. Such integration requires standardization of the interfaces between individual systems and standardization of functions and services. However, this will limit the development possibilities of transport telematics. The solution is a gradual standardization system, that is compatible at individual stages. The benefits of transport telematics systems cannot be precisely determined until the impact and effects of individual systems are sufficiently recognized. A distinction must be made between the benefits associated with implementing telematics systems for transport, the environment, the economy, and society, as well as the benefits in specific situations.

The systematic introduction of telematics techniques will lead to the consideration of telematics systems for specific types of transport. This would reduce the currently dominant focus on only one transport system in transport planning. One of the most important obstacles to the further development of transport telematics is the technical integration of different systems. This challenge arises from the rapid pace of innovation and, to a large extent, the lack of standards.

Another issue with the further development of transport telematics is the typically lengthy implementation process. The time it takes to put the system into practice is often longer than the cycle of developing a new technology. By the time the system is mature enough for practical use, new technologies are already available. However, in the development of telematics, one cannot abandon techniques that have proven effective, even when new solutions emerge. It is important to note that in many areas, outdated but proven technologies have been utilized successfully for an extended period because transitioning to new technologies would be too costly (e.g., securing flights or train traffic).

References

1. Jajszczyk A.: *Podstawy telekomutacji.* WKiŁ, Warszawa 2002.
2. Wawrzyński W., Siergiejczyk M. i inni: Sprawozdanie końcowe grant KBN 5T12C 066 25. Metody wykorzystania środków telematyki we wspomaganiu realizacji zadań transportowych. *Warszawa styczeń*, 2007.
3. Klein L.A.: *Sensor Technologies and Date Requirements if ITS.* Artech Hause, ITS Library, 2001.
4. Siergiejczyk M.: Efektywność eksploatacyjna systemów telematyki transportu. *Prace Naukowe Politechniki Warszawskiej, seria Transport*, nr 67, Oficyna Wydawnicza Politechniki Warszawskiej. Warszawa, 2009.
5. Kołodziejska A., Krzykowska, K. Siergiejczyk M.: Wybrane aspekty wdrożenia usług ITS na drodze szybkiego ruchu, w: Prace Naukowe Politechniki Warszawskiej. *Transport*, nr 113, 2016.
6. Wydro K. B.: *Analiza przepływu informacji w systemach inteligentnego transportu. Praca badawczo – rozwojowa.* Wyd. Instytutu Łączności, Warszawa, 2003.

7. Siergiejczyk M., Gago S.: Platforma teleinformatyczna dla transmisji informacji w transporcie towarowym. *Logistyka*, nr 4, 2008.
8. Campolo, C.: *5G Network Slicing for Vehicle-to-Everything Services*. ieeexplore.ieee.org. 2017.
9. Šeremet, I., Samir, Č.: *Benefits of Using 5G Network Slicing to Implement Vehicle-to-Everything (V2X) Technology*. ieeexplore.ieee.org/2019.
10. Sheri DeTomasi: 5G flexible numerology – What i sit? Why should you care?, community.keysight.com/blogs/next-generation-wireless-communications/blog/2018.
11. Morgan, Y. L.: 2010. Managing DSRC and WAVE standards operations in a V2V scenario. www.researchgate.net/publication/_Managing_DSRC_and_WAVE_standards_operations_in_a_V2V_scenario. 2010.
12. Siergiejczyk M., Krzykowska-Piotrowska K., Olearnik P.: Companion robot communication with road infrastructure as part of IoRT, *Internet Technology Letters*, 2024. DOI:10.1002/itl2.500
13. Carritech.: *Milimetre Wave Spectrum (MMWave) for 5G*. www.carritech.com/news/millimetre-wave-spectrum-mmwaves 2017.
14. Naik G., Choudhury B., Jung-Min P.: *IEEE 802.11bd & 5G NR V2X: Evolution of Radio Access Technologies for V2X Communications*. IEEE, ieeexplore, 2019.
15. Krzykowska-Piotrowska K., Dudek E., Siergiejczyk M., Rosiński A., Wawrzyński W.: Is Secure Communication in the R2I (Robot-to-Infrastructure) Model Possible? Identification of Threats, *Energies*, vol. 14, nr 15, 2021, s. 4702. DOI:10.3390/en14154702
16. Siergiejczyk M., Krzykowska-Piotrowska K.: On the Navigation, Positioning and Wireless Communication of the Companion Robot in Outdoor Conditions, *Energies*, vol. 15, 2022, s. 1–4. DOI:10.3390/en15144936
17. Gago S.: *Teleinformatyka w polskim kolejnictwie. Problemy Kolejnictwa*. Zeszyt 179.Warszawa 2018.
18. Siergiejczyk M., Gago S.: Sieć teleinformatyczna dla potrzeb transportu kolejowego. Materiały konferencyjne V Międzynarodowa Konferencja „Najnowsze technologie w transporcie szynowym". Warszawa, 2016.
19. UIC Future Railway Mobile Communication System. *FRMCS – Transition Technical Guideline for the Application of FRMCS to 4G Networks*, 2024
20. UIC Leaflet No. 407-1 Standardised data exchange between infrastructures for international trains. 2009.
21. COMMISSION REGULATION (EU) No 1305/2014 of 11 December 2014 on the technical specification for interoperability relating to the subsystem "telematics applications for freight services" of the rail system in the European Union and repealing Regulation (EC) No 62/2006 (TAF TSI).
22. COMMISSION REGULATION (EU) No 1273/2013 of 6 December 2013 amending Regulation (EU) No 454/2011 on the technical specification for interoperability relating to the subsystem "telematics applications for passenger services" of the trans-European rail system (TAP TSI).

Chapter 10

Optimising ship routes with capability plots and Dijkstra's algorithm

A case study on anchorage navigation

Jakub Wnorowski and Andrzej Łebkowski

10.1 Introduction

The advancement of modern navigation, particularly in the optimisation of ship routing, has consistently garnered interest from researchers and practitioners in maritime transport [1]. Contemporary navigation systems must address challenges arising from the increasing density of maritime traffic, the growing demand for efficient shipping operations, and the complex interplay between safety considerations and environmental protection. In response, scientists and engineers have developed methods and algorithms to determine ship passage routes effectively, incorporating various factors such as weather conditions, vessel traffic, water depth, and environmental constraints.

The earliest attempts to utilise computers for ship route determination can be traced back to the 1950s and 1960s. During this period, George Hanssen authored a paper proposing a specialised program aimed at enhancing the safety of naval vessels by determining passage routes based on predictable hydrometeorological conditions [2]. With the advent of the first computers capable of performing analytical calculations, weather charts, ocean current charts, and wind charts were employed to map ship routes. Additionally, authors such as those in [3,4] introduced methods that incorporated maps detailing wave strength across specific regions. Subsequent technological advancements enabled the development of more sophisticated techniques and algorithms.

Three primary types of algorithms, along with their various adaptations, are prevalent in contemporary literature concerning the determination of ship passage routes. These include genetic algorithms, graph search algorithms, and swarm algorithms.

Genetic algorithms, inspired by biological processes, incorporate principles such as natural selection, recombination, and mutation. An analysis of the literature on this class of algorithms reveals several key areas of application:

- **Finding a Safe Route in Areas with Limited Manoeuvrability:** One notable application of genetic algorithms is in determining routes in regions with restricted navigation, such as areas with islands and reefs. An example is provided in [5], where the authors demonstrated that traditional algorithms, including ant colony algorithms, performed inadequately in such contexts. To address this limitation, they proposed a genetic algorithm that utilised environmental information to optimise route planning. Another approach was presented in [6], where the authors developed a parallel genetic algorithm for route determination in constrained waters. Their method not only considered the ship's navigational constraints but also incorporated additional criteria such as water depth and travel time to enhance route efficiency and safety.

DOI: 10.1201/9781003686088-10

- **Determining Routes with Criteria Such as Minimising Energy Consumption and Travel Time:** The authors of publications [7,8] addressed this application by presenting an algorithm that integrates elements of particle swarm optimisation and genetic algorithms to determine optimal ship routes. This hybrid approach effectively minimised fuel consumption while considering additional criteria such as travel time, demonstrating its utility in achieving energy-efficient navigation.
- **Determination of Collision Avoidance Routes:** This category focuses on algorithms designed not to determine a complete route from a starting point to an endpoint but rather to adapt local segments of the route to changing conditions around the ship, such as the need to perform collision avoidance manoeuvres. Examples of studies related to this type of algorithm include [9–11]. The authors of these publications have proposed various approaches to collision avoidance, addressing scenarios ranging from harbour operations to applications for autonomous vessels. These algorithms are specifically tailored to ensure the safety of navigation in dynamic environments by making real-time adjustments to the ship's course based on immediate situational requirements.

Graph search algorithms are another class of methods employed in determining ship passage routes. These algorithms are designed to search, analyse, and navigate through graph data structures, which consist of nodes (representing points) and edges (representing connections between points). Basic techniques such as Breadth-First Search (BFS) and Depth-First Search (DFS) are utilised in scenarios where the speed of route generation is prioritised over its optimisation. For instance, the authors of publication [12] used these methods to calculate routes for small vessels. Their findings demonstrated that BFS and DFS algorithms produced routes more quickly than genetic algorithms; however, the routes were less precise.

More advanced graph search algorithms, such as A-star (A*), D-star (D*), and Dijkstra's algorithm, are commonly applied in ship routing. A key challenge in using these methods is defining the transition weights between individual nodes. In classical approaches, the transition weight is often based on the distance between points, optimising the route based on distance. This approach has been applied by Jang and Kim, Pennino et al., Walther et al., and Zhang et al. [13–16].

Alternative methods for defining transition weights have also been explored. For example, the authors of [17,18] utilised Automatic Identification System (AIS) data to construct graphs. In [18], the transition weights were inversely proportional to the number of ships passing between the vertices, while in [17], Delaunay triangulation analysis was used to compute the weights. Another innovative application of graph search algorithms was presented in publication [19], where neural networks were integrated to predict fuel consumption and determine routes that minimise energy use. In addition to focusing on the shortest or least congested paths, some studies have incorporated hydrometeorological conditions into the route optimisation process. For example, publications [20–23] used the A* algorithm in conjunction with weather maps to calculate safe passage routes that account for factors such as wave height, wind speed, and ocean currents. These approaches highlight the versatility of graph search algorithms in addressing various challenges in maritime navigation.

Swarm algorithms represent another widely used category of optimisation techniques for determining ship passage routes. These algorithms are inspired by the collective behaviours observed in nature among animals such as insects, birds, and fish. In natural systems, group interactions—forming swarms, flocks, or schools—often enable the resolution of complex problems through coordinated actions. Swarm algorithms incorporate similar principles and include the following types:

- **Ant Colony Algorithms:** Inspired by the way ants find the shortest path to a food source, these algorithms employ agents (virtual ants) to explore the solution space. As the agents move, they leave virtual pheromone trails, which guide other agents towards optimal solutions.
- **Particle Swarm Algorithms:** Based on the behaviour of schools of fish or flocks of birds, these algorithms use particles (representing points in the solution space) that traverse the search space. Each particle is influenced by its own past experiences and the collective knowledge of the swarm, allowing it to converge on the best solution.
- **Bee-Inspired Algorithms:** These mimic the behaviour of bees foraging for food. Bees identify high-quality food sources and recruit other bees to exploit these resources, enabling the algorithm to focus on promising areas in the solution space.
- **Firefly Algorithms:** The core principle of these algorithms is the attraction between fireflies, which is based on their brightness—a metaphor for the quality of a solution. The brightness of each firefly corresponds to the value of the objective function and diminishes with distance. Fireflies with lower brightness are attracted to brighter ones, resulting in iterative improvements to the solution. The closer two fireflies are, the stronger the attraction, fostering convergence on optimal solutions.

In the literature on swarm algorithms, all the aforementioned types have been applied to ship passage route determination. One example is presented in publication [24], where the authors demonstrated how ant-inspired algorithms could enhance the efficiency of maritime navigation. They integrated the Electronic Chart Display and Information System (ECDIS) with a Geographic Information System (GIS) platform to model the maritime space.

The algorithm simulated the behaviour of ants searching for the optimal path to a food source, where the "ants" (representing ships) explored the search space to identify the safest route. Experimental results showed that the routes generated using the ant algorithm were more efficient than those determined by the great circle method. However, Zhang et al. [25] offered a different perspective on ant algorithms, highlighting certain limitations. The authors pointed out that classical ant algorithms often suffer from drawbacks such as a tendency to get trapped in local minima and slow convergence speeds, which negatively impact both the quality and efficiency of route determination. Beyond ship passage route determination, ant algorithms have been applied in various other fields, including:

- **Route planning for mobile robots** [26,27].
- **Passage route determination for units involved in maritime rescue missions** [28,29].
- **Collision avoidance route planning** [30].

These diverse applications underscore the versatility of ant-inspired algorithms, despite their limitations, in solving complex routing and navigation challenges.

In contrast to ant algorithms, particle swarm algorithms are known for their faster convergence rates and their ability to optimise multiple variables simultaneously. This advantage was demonstrated in [31], where the authors combined the particle swarm algorithm with Dijkstra's algorithm to determine the shortest route between starting and ending points while accounting for obstacles. The algorithm utilised a cost function based on the total length of the route, enabling it to efficiently generate optimised paths. Another notable example of particle swarm algorithm application in ship route planning is presented in Ref. [32]. In this study, the authors introduced the concept of chaos multi-population particle swarm optimisation (CMPSO), an enhanced version of the particle swarm algorithm incorporating chaos theory. This improvement allowed for better handling of complex optimisation problems, making the approach particularly effective in scenarios with dynamic or unpredictable conditions.

Bee-inspired swarm algorithms have found applications in route optimisation, particularly within maritime navigation. These algorithms are based on mechanisms observed in bees, such as their ability to search for and communicate information about food sources. This behaviour provides bee algorithms with a strong capacity for exploring the solution space and effectively avoiding local minima.

In maritime navigation, bee-inspired algorithms are utilised for dynamic adaptation to changing conditions, such as obstacles along the route or variations in weather. These capabilities make them valuable tools for ensuring safety and efficiency. For instance, Li et al. [33] developed an algorithm using bee-inspired principles to calculate escape routes for submarines, allowing them to avoid detection by hostile vessels. This application highlights the adaptability and effectiveness of bee algorithms in managing complex and dynamic navigational scenarios.

This paper presents the application of a modified Dijkstra's algorithm combined with position-keeping capability plots (CPs) to determine a passage route that minimises the impact of environmental forces on the ship's hull, thereby reducing energy consumption. The method is specifically designed for autonomous vessels equipped with a positioning system based on a set of anchors. For such vessels, the primary challenge lies in navigating from the starting point to the pre-planned anchorage area. Unlike dynamic positioning (DP) systems, where thrusters play a crucial role, the thrusters in these vessels have significantly lower power and, as a result, limited manoeuvrability. Consequently, the passage route must be optimised to minimise the effects of external disturbance factors such as wind, waves, and currents. The proposed algorithm was verified using a simulator developed within the Unity3D environment, demonstrating its effectiveness in addressing the unique navigational challenges of anchor-positioned vessels.

10.2 Mathematical models used in route planning

To establish a ship's passage route using the proposed method, it is imperative to understand how environmental forces interact with the vessel's structure. This interaction is systematically represented through CPs, a tool commonly employed in vessels equipped with DP systems. CPs are used to evaluate the operational capabilities of a ship under various environmental conditions, providing a detailed visualisation of its performance limits. A typical CP (Figure 10.1) includes three critical components: the silhouette of the vessel, numerical annotations denoting the angles of environmental forces relative to the hull, and a boundary line representing the maximum permissible wind speed at specific impact angles. This boundary indicates the ship's ability to maintain its position without exceeding its operational limits. By incorporating CPs into the route determination process, a more precise assessment of the ship's capability to counteract external forces is achieved, paving the way for energy-efficient navigation.

The mathematical models used to calculate these interactions take into account environmental forces originating from wind, ocean currents, and sea waves. These forces are modelled individually and integrated to assess their cumulative effect on the vessel's hull. The equations used are derived from fundamental principles of fluid dynamics and ship stability, tailored to reflect the unique characteristics of the vessel under study.

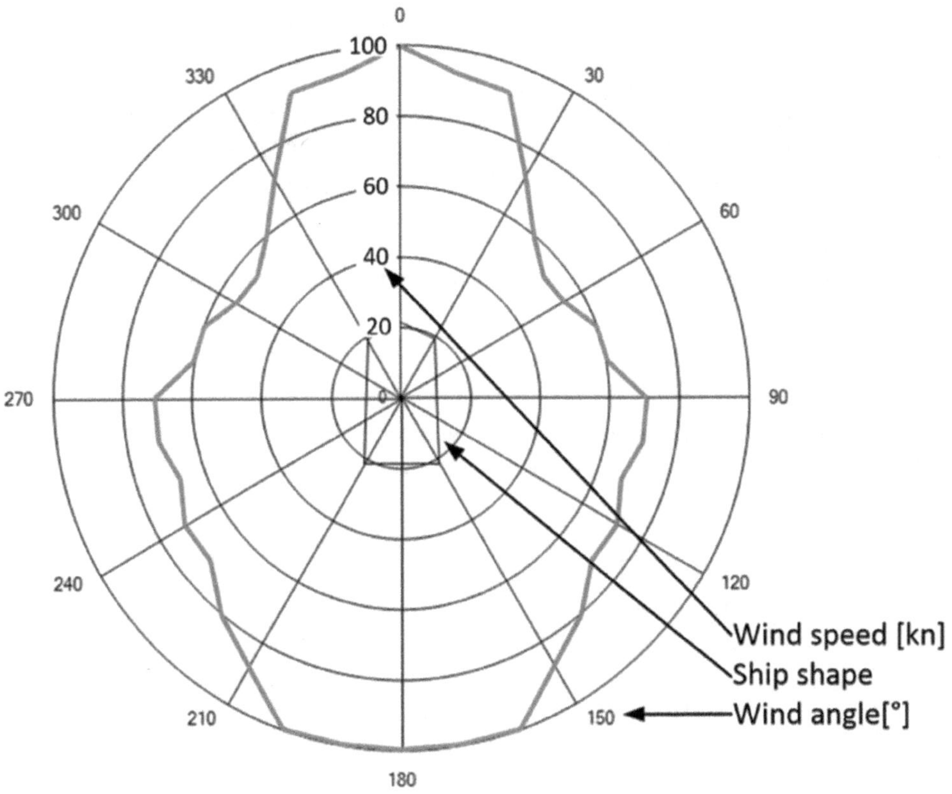

Figure 10.1 An example of the Capability Plot.

This chapter outlines the mathematical frameworks underpinning the analysis, beginning with the forces generated by wind, followed by those arising from ocean currents and sea waves. The models integrate various parameters, including the ship's dimensions, environmental conditions, and hydrodynamic coefficients, to provide an accurate representation of real-world scenarios. By understanding these forces and their interaction with the ship, it becomes possible to develop an optimised passage route that reduces energy consumption and enhances operational safety.

10.2.1 Forces induced by wind

Wind-induced forces exert a significant influence on a vessel's movement and stability, particularly in maritime operations involving DP or precise navigation [34]. To model these effects, it is essential to quantify the forces acting along the longitudinal (X-axis), lateral (Y-axis), and rotational (torque) axes of the ship's hull. These forces can be represented mathematically as follows [35]:

$$Wind_X = q_w \cdot C_x(\varphi_w) \cdot A_{FW} \tag{10.1}$$

$$Wind_Y = q_w \cdot C_Y(\varphi_w) \cdot A_{LW} \tag{10.2}$$

$$Wind_M = q_w \cdot C_M(\varphi_w) \cdot A_{FW} \cdot H_{FW} \tag{10.3}$$

where $Wind_X$ is the wind force acting on the X-axis of the ship [N], q_w is the factor determining the wind pressure on the ship's hull [kg/m-s²], φ_w is the angle of wind impact relative to the $L_{pp}/2$ point [°], C_X is the shape coefficient for the X-axis [−], A_{FW} is the cross-sectional area of the above-water hull [m²], $Wind_Y$ is the wind force acting on the Y-axis of the ship [N]; C_Y is the shape coefficient for the Y-axis [−], A_{LW} is the lateral surface area of the hull above the waterline [m²], $Wind_M$ is the torque generated by wind force [N-m], C_M is the shape coefficient for the torque [−], and H_{FW} is the offset of the centroid of the lateral above-water surface relative to the $L_{pp}/2$ point [m];

The wind pressure factor (q_w) is calculated using the following equation:

$$q_w = \frac{1}{2} \cdot \rho_a \cdot V_W^2 \tag{10.4}$$

where ρ_a is the air density [kg/m³] and V_w is the wind speed [m/s].

The shape coefficients (C_X, C_Y, C_M) are determined based on the ship's hull geometry and are expressed as [35]

$$C_X = -CD_{lAF} \cdot \frac{\cos(\varphi_w)}{1 - \frac{\delta}{2}\left(1 - \frac{CD_l}{CD_t}\right)\sin^2(2\varphi_w)} \tag{10.5}$$

$$C_Y = CD_t \cdot \frac{\sin(\varphi_w)}{1 - \frac{\delta}{2}\left(1 - \frac{CD_l}{CD_t}\right)\sin^2(2\varphi_w)} \tag{10.6}$$

$$C_M = \left[\frac{H_{FW}}{L_{oa}} - 0.18\left(\varphi_w - \frac{\pi}{2}\right)\right] \cdot C_Y(\varphi_w) \tag{10.7}$$

$$CD_l = CD_{lAF}(\varphi_w) \cdot \left(\frac{A_{FW}}{A_{LW}}\right) \tag{10.8}$$

where C_X is the shape coefficient for the X-axis [−], CD_{lAF} is the longitudinal resistance coefficient [−], φ_w is the wind angle [°], δ is the cross-force parameter [−], CD_l is the longitudinal resistance [−], CD_t is the transverse resistance [−], C_Y is the shape coefficient for the Y-axis [−], C_M is the shape coefficient for torque [−], H_{FW} is the offset of the centroid of the lateral above-water surface relative to the $L_{pp}/2$ point [m], L_{oa} is the length over all [m], A_{LW} is the lateral surface area of the hull above the waterline [m²], and A_{FW} is the cross-sectional area of the above-water hull [m²].

To provide a comprehensive understanding of these forces, Figure 10.2 illustrates the variation of hull shape coefficients (C_X, C_Y, C_M) with respect to the angle of wind impact (φ_w). The values of these coefficients are highly sensitive to the shape and dimensions of the hull, highlighting the importance of accurate modelling in ensuring effective navigation.

The inclusion of these equations allows for the precise quantification of wind-induced forces, which is critical for ensuring that a vessel's operational capabilities are maintained under varying environmental conditions.

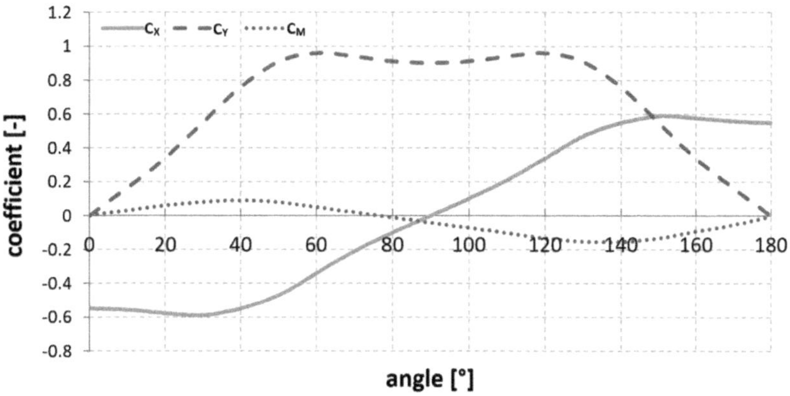

Figure 10.2 Hull shape coefficients.

10.2.2 Forces induced by sea current

Ocean currents significantly influence a ship's stability and trajectory, requiring accurate modelling to ensure effective navigation and energy efficiency. These forces are modelled by considering their effects on the longitudinal (*X*-axis), lateral (*Y*-axis), and rotational (torque) axes of the vessel. The equations representing these forces are given as follows [35]:

$$Current_X = q_c \cdot K_x(\varphi_C) \cdot A_{FC} \tag{10.9}$$

$$Current_Y = q_c \cdot K_Y(\varphi_C) \cdot A_{LC} \tag{10.10}$$

$$Current_M = q_c \cdot K_M(\varphi_C) \cdot A_{FC} \cdot H_{FC} \tag{10.11}$$

where $Current_X$ is the sea current force acting on the X-axis of the ship [N], q_C is the factor determining the sea current pressure on the ship's hull [kg/m-s^2], φ_C is the angle of sea current impact relative to the $L_{pp}/2$ point [°], K_X is the shape coefficient for the *X*-axis [–], A_{FC} is the cross-sectional area of the hull under waterline [m^2], $Current_Y$ is the sea current force acting on the *Y*-axis of the ship [N], K_Y is the shape coefficient for the *Y*-axis [–], A_{LC} is the lateral surface area of the hull under waterline [m^2], $Current_M$ is the torque generated by sea current force [N-m], and H_{FC} is the offset of the centroid of the underwater lateral surface relative to the $L_{pp}/2$ point [m].

The sea current pressure factor (q_C) is calculated using the following equation:

$$q_C = \frac{1}{2} \cdot \rho_C \cdot V_C^2 \tag{10.12}$$

where ρ_C is the air density [kg/m^3], and V_c is the sea current speed [m/s].

The shape coefficients (K_X, K_Y, K_M) are essential for accurately quantifying the interaction between ocean currents and the ship's hull. These coefficients are analogous to those used in wind force modelling but are adapted to account for the submerged portions of the vessel. The determination of K_X, K_Y, K_M relies on the longitudinal and transverse underwater surface areas, ensuring that the model reflects the unique hydrodynamic characteristics of the vessel.

10.2.3 Forces induced by sea waves

Sea waves exert significant forces on a ship's hull, influencing its stability, propulsion, and manoeuvrability. To model these effects, it is essential to account for the interaction between wave dynamics and the ship's structure. These forces are determined based on widely recognised wave spectra, such as the JONSWAP [36] spectrum and the Pierson-Moskowitz spectrum [36], the latter being commonly employed by classification societies like Det Norske Veritas (DNV). This section presents the mathematical models used to determine wave-induced forces [37].

$$X_{waves} = q \cdot B \cdot h\left(Dir, bow_a, C_{WLaft}\right) \cdot f\left(T_s'\right) \tag{10.13}$$

$$h\left(Dir, bow_a, C_{WLaft}\right) = 0.09 \cdot h_1\left(Dir, bow_a, C_{WLaft}\right) \cdot h_2(Dir) \tag{10.14}$$

$$h_{1A}\left(bow_a\right) = 0.8 \cdot bow_a^{.45} \tag{10.15}$$

$$h_{1B}\left(C_{WLaft}\right) = 0.7 \cdot C_{WLaft}^2, C_{WLaft} \in \left[0.85, 1.15\right] \tag{10.16}$$

$$dir(Dir) = \begin{cases} Dir, 0 \le Dir \le \pi \\ 2\pi - Dir, \pi \le Dir \le 2\pi \end{cases} \tag{10.17}$$

$$h_1\left(Dir, bow_a, C_{WLaft}\right) = h_{1A}\left(bow_a\right) + \frac{dir(Dir)}{\pi}\left(h_{1B}\left(C_{WLaft}\right) - h_{1A}\left(bow_a\right)\right) \tag{10.18}$$

$$h_2(Dir) = 0.05 + 0.95 \cdot \tan^{-1}\left(1.45 \cdot (dir(Dir) - 1.75)\right) \tag{10.19}$$

$$f\left(t'\right) = \begin{cases} 1, \text{ if } t' < 1 \\ t'^{-3} \cdot e^{1-t'^{-3}}, \text{ if } t' \ge 1 \end{cases} \tag{10.20}$$

$$Y_{waves} = q \cdot L_{OS} \cdot (0.09 \cdot \sin(Dir)) \cdot f\left(T_{sw}'\right) \tag{10.21}$$

$$M_{waves} = Y_{waves} \cdot (X_{Los} + \left(0.05 - 0.14 \cdot \frac{dir(Dir)}{\pi}\right) \cdot L_{OS} \tag{10.22}$$

$$T_s' = \frac{T_z}{0.9 \cdot L_{pp}^{0.33}} \tag{10.23}$$

$$T_{sw}' = \frac{T_z}{0.75 \cdot B^{0.5}} \tag{10.24}$$

$$H_s = 0.3125 \cdot V_w - 0.62 \tag{10.25}$$

$$T_z = 0.741 \cdot V_w + 0.536 \tag{10.26}$$

where X_{waves} is the sea waves' force acting on the X-axis of the ship [N], q is the water pressure on the ship's hull [kg/m·s²], B is the maximum breadth at water line [m], Dir is the waves coming from direction [°], bow_{α} is the angle between the vessel X-axis and a line drawn from the foremost point in the water line to the point at $y=B/4$ on the water line [°], $C_{\underline{WLaft}}$ is the water plane area coefficient of the water plane area behind the midship [−], Y_{waves} is the sea waves' force acting on the Y-axis of the ship [N], L_{os} is the longitudinal distance between the foremost and aftmost point under water [m], M_{waves} is the torque generated by sea current force [N·m], X_{Los} is the longitudinal position of $L_{os}/2$ [m], L_{pp} is the length between perpendiculars [m], H_s is the significant wave height [m], A_{WLaft} is the water plane area for $x<0$ [m²], T_c' is the sea wave period for surge [s], T_{sw}' is the sea wave period for sway [s], T_Z is the zero crossing periods [s], and V_W is the actual wind speed [m/s].

10.2.4 Technical parameters of the vessel

To accurately model the environmental forces acting on a vessel and to implement the proposed route optimisation algorithm, it is essential to consider the ship's technical characteristics. These include its physical dimensions, hydrodynamic properties, and propulsion system specifications. This section outlines the key parameters of the vessel used in this study.

The ship's primary dimensions and hydrodynamic properties are summarised in Table 10.1. These parameters are critical for calculating forces due to wind, currents, and waves, as well as for generating CPs that define the vessel's operational limits.

The ship's propulsion system consists of four thrusters: a bow thruster, a stern thruster, and two main propellers. Table 10.2 provides detailed information about their thrust capabilities and power requirements.

Table 10.1 Technical specifications of the vessel

Parameter	*Value*
Length overall (L_{oa}) [m]:	72.7
Length between perpendiculars (L_{pp}) [m]:	64
Breadth [m]:	11.6
Draught [m]:	3.4
Displacement [T]:	1886
Distance between the foremost and aftmost points of the hull below the surface at design draft, even keel [m]:	67.1
Water plane area [m²]:	639
Projected longitudinal area above water [m²]:	437
Surge position of the geometric centre of the projected longitudinal area above water with respect to $L_{pp}/2$ [m]:	0.1
Projected longitudinal area below water [m²]:	223
Surge position of geometric centre of the projected longitudinal area below water with respect to $L_{pp}/2$ [m]:	−2.9
Surge position of water line centre with respect to $L_{pp}/2$ [m]:	−1.5
Projected transverse area above water [m²]:	140
Projected transverse area below water [m²]:	36
Fuel consumption at 100% main engine load [l/h]	246
Propulsion system efficiency [%]	41

Table 10.2 Thruster specification

Thruster	*Thrust max [kN]*	*Thrust min [kN]*	*Power [kW]*
Bow thruster	94	−94	447
Stern thruster	94	−94	447
Port propeller	179	−179	2,000
Stbd propeller	179	−179	2,000

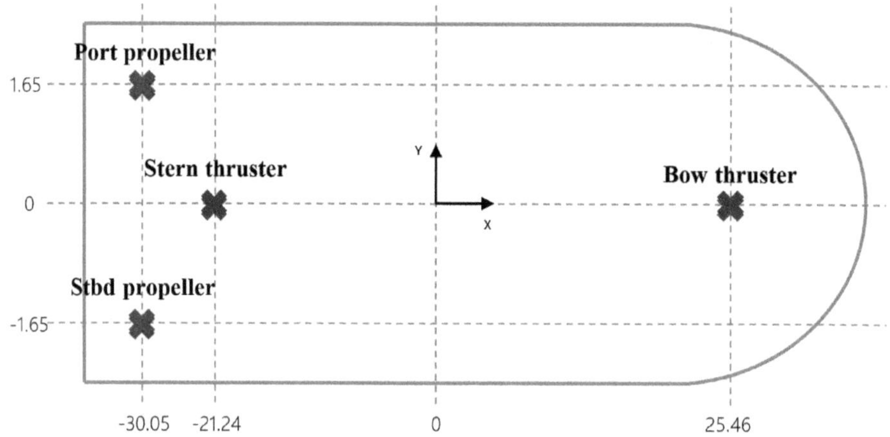

Figure 10.3 Diagram of Thruster Arrangement on the Ship.

The positioning of the thrusters is illustrated in Figure 10.3, showing their arrangement along the length of the vessel. This configuration allows for precise manoeuvrability, which is essential for efficient navigation in complex maritime environments.

The creation of CPs involves analysing the environmental forces acting on the ship's hull at various angles and determining the vessel's ability to counteract these forces using its propulsion system. This process begins with evaluating the forces generated by wind, currents, and waves, starting from an initial angle of 0°. The forces are incrementally adjusted for a range of angles to determine the ship's operational limits. For each angle, the maximum wind speed at which the vessel can maintain its position is calculated. This is achieved by distributing the required thrust among the ship's thrusters using an optimisation method, such as Johansen's allocation formulation [38]. If the vessel cannot maintain its position at a given wind speed, the speed is gradually reduced until the required balance is achieved. The final wind speed for each angle is recorded, and the process is repeated for subsequent angles, typically in increments of 10°. Once the analysis is complete, the results are plotted to produce the CP, as illustrated in Figure 10.4. The plot includes a boundary line that connects the maximum wind speed values for all tested angles, providing a visual representation of the vessel's operational capability under varying environmental conditions. The CP also incorporates the ship's geometric parameters, propulsion limits, and hydrodynamic characteristics to ensure accuracy.

The resulting CP enables precise evaluation of the ship's performance in maintaining its position against external forces. This information is integral to the proposed route optimisation algorithm, allowing the identification of routes that minimise environmental impact while staying within the vessel's operational constraints.

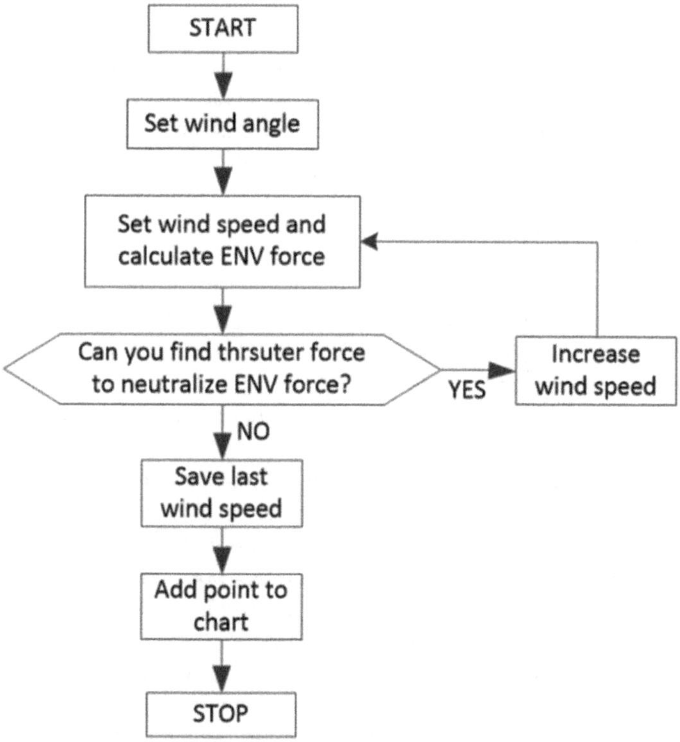

Figure 10.4 CP's algorithm.

10.3 Optimising ship routes using capability plots and Dijkstra's algorithm

One of the initial and most critical steps in maritime operations is determining the optimal route for a vessel between two geographical points, such as the current position of a ship and its designated anchorage. The growing complexity of maritime traffic and the need for energy-efficient and environmentally sustainable solutions have highlighted the importance of developing advanced routing algorithms.

Traditional navigation methods often involve plotting direct courses, which may not adequately address dynamically changing environmental conditions such as wind, waves, and ocean currents. These forces can significantly impact a ship's stability, energy consumption, and operational safety. Modern computational approaches, which utilise real-time data and adaptive algorithms, offer a way to overcome these limitations.

This chapter presents a novel approach to route planning that combines CPs with a modified version of Dijkstra's algorithm. The CPs are employed to evaluate the ship's capability to maintain position under varying environmental forces, while the algorithm determines the path with the lowest cumulative transition weight between waypoints. The navigational area is represented as a grid of waypoints, and each transition between points is weighted based on the environmental forces affecting the ship's hull. The chapter details the theoretical framework behind this approach, the implementation of the algorithm, and the criteria used to evaluate its effectiveness. Examples are provided to demonstrate how the algorithm performs under different environmental scenarios, highlighting its ability to reduce energy consumption and improve operational efficiency.

10.3.1 The algorithm's concept

The developed algorithm is designed to determine the optimal route for a vessel equipped with multiple anchors, guiding it from its current location to the designated site for deploying the first anchor. Figure 10.5 illustrates a segment of the anchor planning tool, alongside the navigation scenario used to demonstrate the algorithm's functionality.

The diagram above identifies two key points: A, representing the starting location, and B, denoting the endpoint. It is assumed that the planned route should terminate at the site designated for the initial anchor deployment. The distance separating these points is 3.13 nautical miles (NM).

The initial step in determining the vessel's route involves defining the manoeuvrable area within which the ship can operate. For this defined area, a grid of waypoints is generated, with the density of these points specified by the user. Figure 10.6 illustrates the navigational scenario depicted in Figure 10.5, now incorporating the waypoint grid. Each waypoint within the grid is characterised by its geographic coordinates, including latitude and longitude.

In preparation for the anchoring operation, the ship's crew first arrives at the designated anchorage and conducts an initial assessment of the area. This evaluation precedes the analysis of anchor placement and manoeuvring strategies. Consequently, the algorithm begins route planning from the waypoint closest to the vessel's current position.

To demonstrate the functionality of the algorithm, a waypoint density of 10 was selected, resulting in a grid where each row and column contains 11 points (with numbering starting from 0). It was further assumed that the wind acting on the vessel has a speed of 10 m/s and a direction of 90°.

The route determination in this scenario is a classic graph search problem. A commonly used approach to solve such problems is Dijkstra's algorithm. This algorithm identifies the shortest path between two vertices in a weighted graph, where each edge is assigned a numerical value known as a weight. As a greedy algorithm, Dijkstra's method makes locally optimal decisions at each step, ultimately arriving at a globally optimal solution. However, for the algorithm to perform correctly, the graph must not contain edges with negative weights.

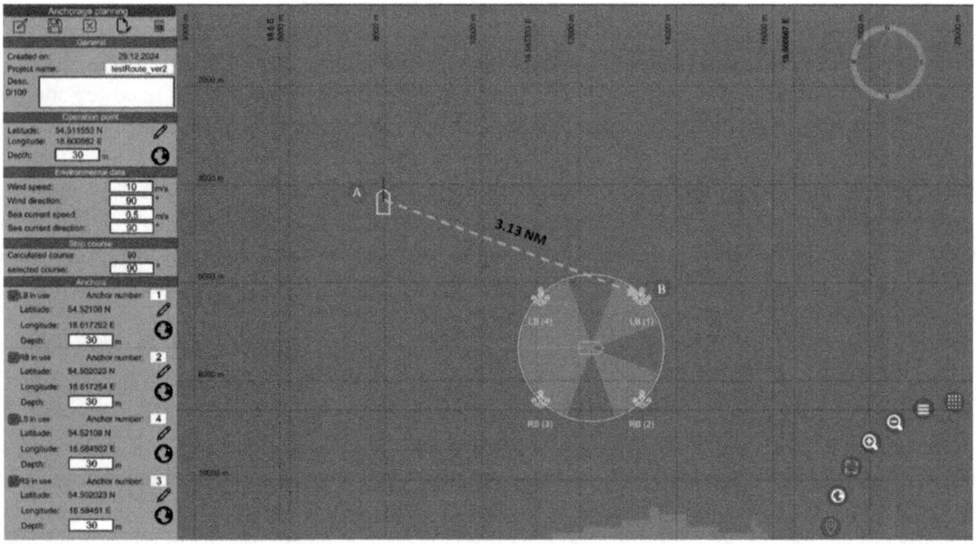

Figure 10.5 Navigation scenario.

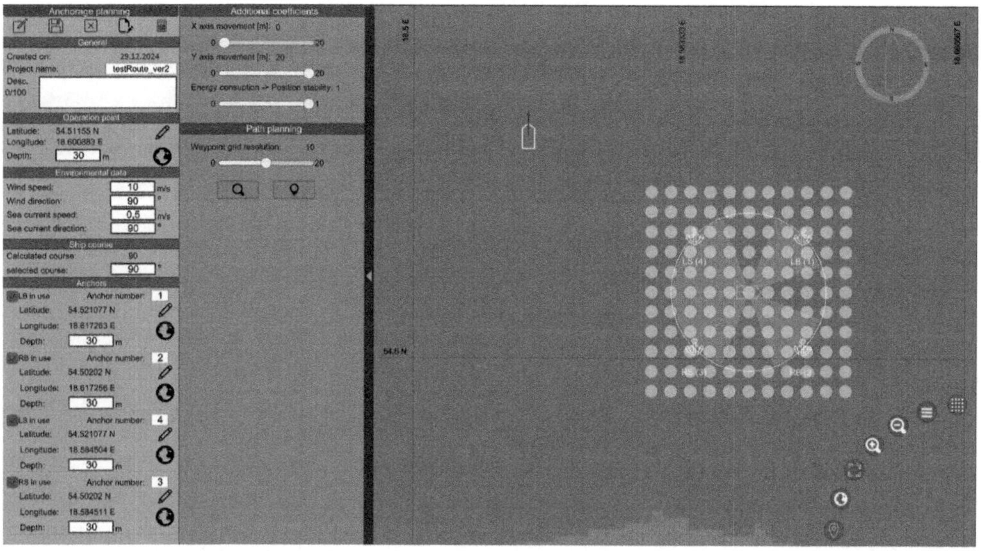

Figure 10.6 Navigational situation with the waypoint grid.

The transition weights between individual points in the grid are calculated by converting the vessel's CP into a set of coefficients that account for the wind's angle of attack on the ship's hull. This conversion process is carried out using Equation (10.27).

$$\frac{F\alpha(\alpha)}{F_{\max}} \tag{10.27}$$

where $F_{\alpha}(\alpha)$ is the resultant environmental force currently acting on the ship's hull at angle α; $F_{\max}$ is the maximum resultant environmental force that can act on the ship's hull.

The maximum environmental force that the ship's hull can withstand is established during the development of the CP. For the vessel described in Section 2, this maximum environmental force is calculated to be 320 kN. Figure 10.7 presents the transition weights assigned to individual points in the grid, which vary depending on the wind's angle of impact on the hull.

The next step, after defining the table presented in Figure 10.7, involves creating the transition matrix that defines the relationships between individual points. This matrix specifies the connections between each point and its neighbouring points, as well as the associated cost of transitioning to each of them. To establish the relationships between points (i.e., from each point to its neighbouring points), a matrix is constructed where the rows correspond to the current point being evaluated by the algorithm, and the columns represent its potential neighbours. For the given scenario, the matrix dimensions are as follows:

[number of waypoints × number of waypoints]

For the navigational scenario depicted in Figure 10.5, the transition matrix has dimensions of [121 × 121]. The specific points within the matrix are organised based on the order illustrated in Figure 10.8.

$F_\alpha(\alpha)/F_{max}$

wind angle [°]	weight	wind angle [°]	weight
0	0.0533	**110 / 250**	0.3096
10 / 350	0.0850	**120 / 240**	0.2778
20 / 340	0.1165	**130 / 230**	0.2746
30 / 330	0.1762	**140 / 220**	0.2133
40 / 320	0.2113	**150 / 210**	0.1786
50 / 310	0.2448	**160 / 200**	0.1205
60 / 300	0.2776	**170 / 190**	0.0903
70 / 290	0.3095	**180**	0.0611
80 / 280	0.3115		
90 / 270	0.3121		
100 / 260	0.3113		

Figure 10.7 Conversion of the capability plot to a transition weight matrix.

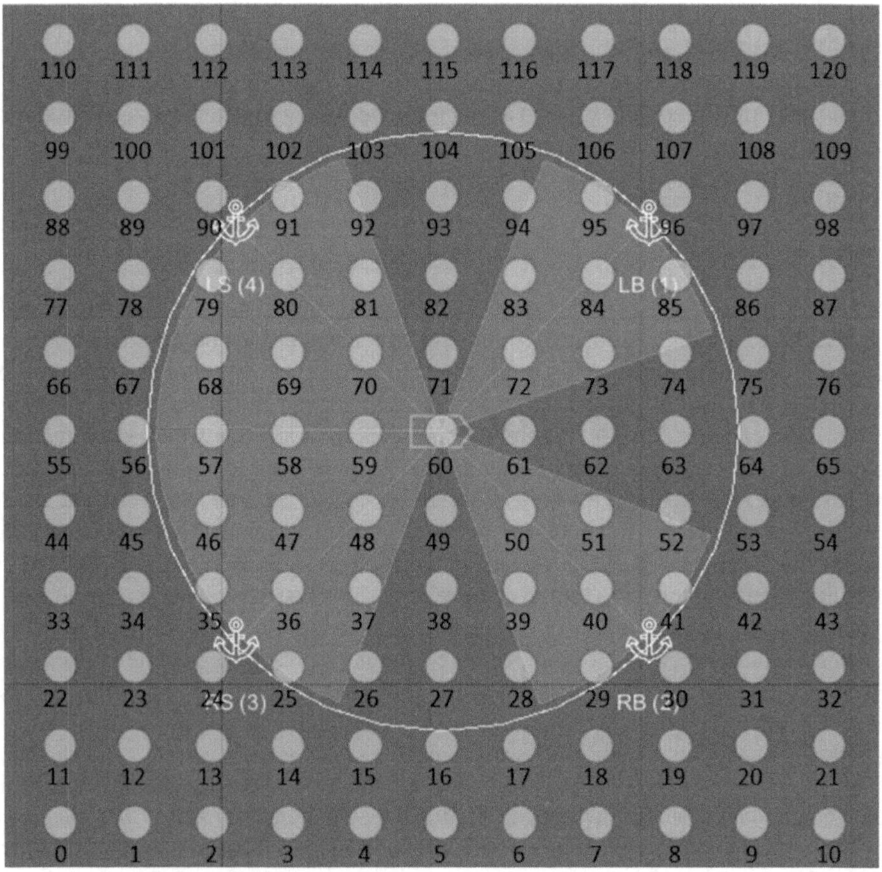

Figure 10.8 Numbering of individual points in the waypoint grid.

At the outset of table initialisation, all transition weights are assigned a default value of 10, indicating that no passage exists between the respective points. Subsequently, for each point, its neighbouring points are identified, the transition weights between them are calculated, and these values are entered into the corresponding positions within the transition matrix.

A fragment of the table containing the completed transition weights is presented in Figure 10.9. To accommodate space constraints, the numerical values in the table have been rounded to two decimal places. The full, unrounded values can be referenced in Figure 10.7.

10.3.2 Determining the ship's route without using an algorithm

From the navigational scenario depicted in Figure 10.5, it is evident that the distance between the ship's current position and the anchoring point is 3.13 NM (approximately 5.8 km). For such a short distance, manual control with a steady course can be employed to navigate to the endpoint, as shown in Figure 10.10.

To evaluate the impact of environmental forces acting on a ship travelling along a course of 110° over a distance of 5800 m, a simplified coefficient was introduced to facilitate subsequent comparisons with the route generated by the algorithm. This coefficient is defined as follows:

$$T = \sum_{i=1}^{n} d_i \cdot k_i \tag{10.28}$$

where T is the route cost function [−], n is the number of transitions between waypoints [−], i is the current transition between waypoints [−], d is the distance between waypoints [m], and k is the cost of transition between waypoints [−]

In Equation (10.28), the transition cost between waypoints is derived from the table presented in Figure 10.6.

For the navigational scenario illustrated in Figure 10.10, substituting the distance values between points and the transition costs from Figure 7—while considering the wind direction and the ship's course—resulted in a route cost function value of 687. The energy consumption for navigating the route shown in Figure 10.10 is calculated using the thruster data provided in

	Next Waypoint																	
Actual Waypoint	0	1	2	3	4	5	6	7	8	9	10	11	12	13	14	15	16	17
0	0	0.30	10	10	10	10	10	10	10	10	10	0.05	0.19	10	10	10	10	10
1	0.30	0	0.30	10	10	10	10	10	10	10	10	0.7	0.05	0.19	10	10	10	10
2	10	0.30	0	0.30	10	10	10	10	10	10	10	10	0.7	0.05	0.19	10	10	10
3	10	10	0.30	0	0.30	10	10	10	10	10	10	10	10	0.7	0.05	0.19	10	10
4	10	10	10	0.30	0	0.30	10	10	10	10	10	10	10	10	0.7	0.05	0.19	10
5	10	10	10	10	0.30	0	0.30	10	10	10	10	10	10	10	10	0.7	0.05	0.19
6	10	10	10	10	10	0.30	0	0.30	10	10	10	10	10	10	10	10	0.7	0.05
7	10	10	10	10	10	10	0.30	0	0.30	10	10	10	10	10	10	10	10	0.47
8	10	10	10	10	10	10	10	0.30	0	0.30	10	10	10	10	10	10	10	10
9	10	10	10	10	10	10	10	10	0.30	0	0.30	10	10	10	10	10	10	10
10	10	10	10	10	10	10	10	10	10	0.30	0	10	10	10	10	10	10	10
11	0.12	0.29	10	10	10	10	10	10	10	10	10	0	0.30	10	10	10	10	10
12	0.14	0.12	0.29	10	10	10	10	10	10	10	10	0.30	0	0.30	10	10	10	10
13	10	0.14	0.12	0.29	10	10	10	10	10	10	10	10	0.30	0	0.30	10	10	10
14	10	10	0.14	0.12	0.29	10	10	10	10	10	10	10	10	0.30	0	0.30	10	10
15	10	10	10	0.14	0.12	0.29	10	10	10	10	10	10	10	10	0.30	0	0.30	10
16	10	10	10	10	0.14	0.12	0.29	10	10	10	10	10	10	10	10	0.30	0	0.30
17	10	10	10	10	10	0.14	0.12	0.29	10	10	10	10	10	10	10	10	0.30	0

Figure 10.9 The fragment of the table with transition weights.

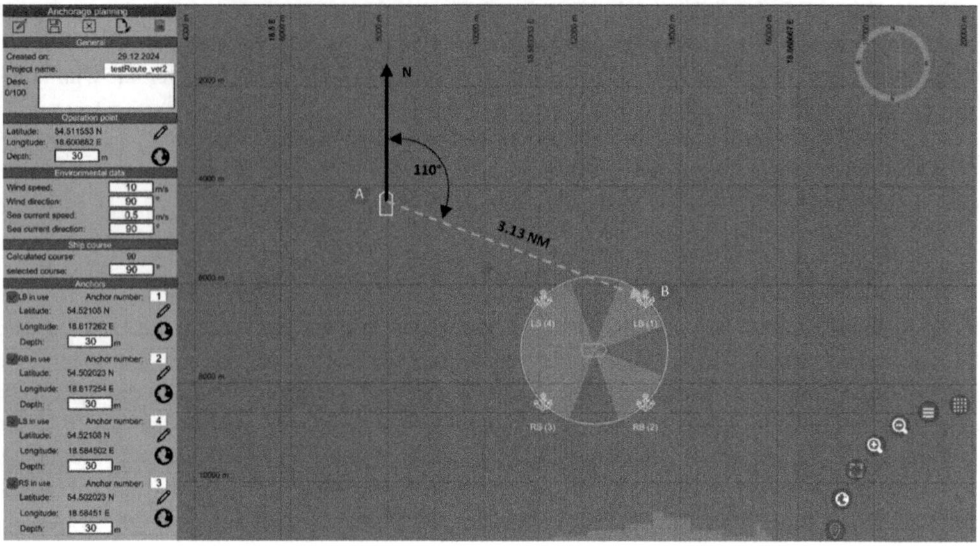

Figure 10.10 Ship route in case of manual control.

Table 10.3 The total energy consumption

Phase	*S [NM]*	*V [kn]*	*t [h]*	*Course [°]*	*True Wind Angle [°]*	*Env. Force Vector [X-axis; Y-axis]*	*Sum of Thrust [kN]*	*Fuel Consumption [l]*	*Energy Consumption [kWh]*
Point A -> Point B	3.13	4	0.782	110	90	[20 kN; 33 kN]	42	424	3745.7

Table 10.2 and the ship specifications listed in Table 10.1. The total energy consumption values are summarised in Table 10.3.

In Table 10.3,**Phase,** current route segment; ***S,*** distance between points; ***V,*** speed of the ship; ***t,*** the duration of the navigation of the given segment; **Course,** ship's course; **True wind angle,** angle of wind impact on the vessel; **Env. Force vector,** environmental forces affecting the vessel over a given distance; **Sum of thrust,** thrust generated by propellers; **Fuel consumption,** fuel consumption per route section; and **Energy consumption,** energy consumption per route section.

10.3.3 Pathfinding with Dijkstra's algorithm and capability plots

The passage route calculated using Dijkstra's algorithm for the navigational scenario depicted in Figure 10.5 is presented in Figure 10.11.

In the figure above, the starting point of the calculated route is highlighted in red, with the colour gradually transitioning to green as each waypoint approaches the destination.

The cumulative transition weight for the identified route is 2.44, representing the lowest achievable value under the given anchorage parameters. While alternative routes with the same total weight can be found, from the perspective of the algorithm's operation, these routes are considered equivalent.

The route cost function value is provided in Table 10.4.

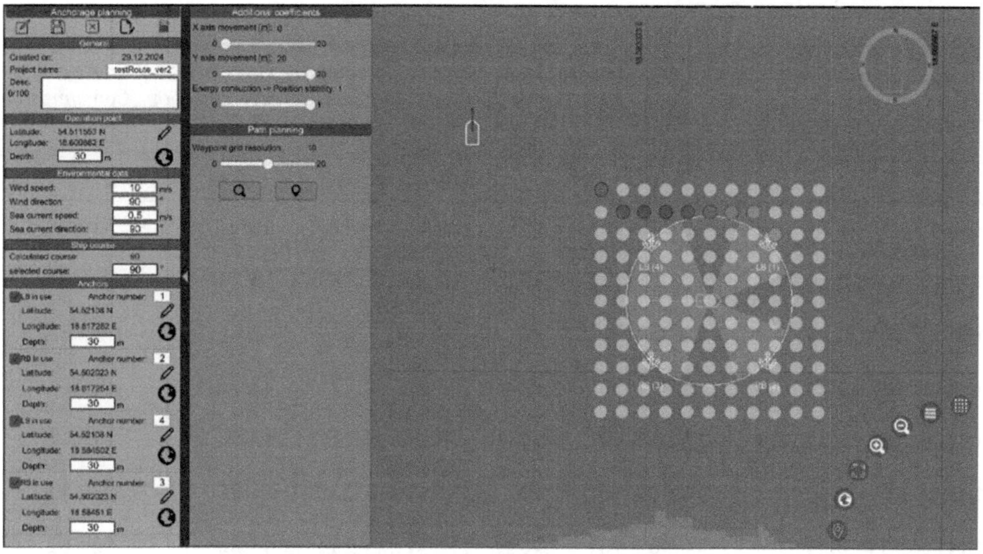

Figure 10.11 The route determined using algorithm.

Table 10.4 The route cost function

Waypoints' Edge	*d [m]*	*Course [°]*	*k [−]*	$d \cdot k$
Point A -> 110	2500	110	0.116	290
110 -> 100	565	135	0.227	128
100 -> 101	400	90	0.053	21.2
101 -> 102	400	90	0.053	21.2
102 -> 103	400	90	0.053	21.2
103 -> 104	400	90	0.053	21.2
104 -> 105	400	90	0.053	21.2
105 -> 106	400	90	0.053	21.2
106-> 96 (Point B)	565	135	0.227	128
			Sum	673.2

For the route determined by the algorithm, the route cost function value was 673.2, which is approximately 3% lower than the cost when the ship follows a direct path to the endpoint. The energy consumption for this case can be calculated similarly to the direct route. Table 10.5 provides an analysis of the energy consumption for the algorithm-determined route, considering the current segments of the route.

The energy consumption along the route determined by the algorithm was 3567.41 kWh, a reduction of approximately 5% compared to the direct route to the end point.

10.3.4 Additional examples of calculated routes

The algorithm described in this publication is capable of determining routes under various hydrometeorological conditions. Presented below are examples of calculated routes, along with details of the specific weather conditions under which they were generated.

Table 10.5 The total energy consumption for the route determined by the algorithm

Phase	*S [NM]*	*V [kn]*	*t [h]*	*Course [°]*	*True Wind Angle [°]*	*Env. Force Vector [X-axis; Y-axis]*	*Sum of Thrust [kN]*	*Fuel Consumption [l]*	*Energy Consumption [kWh]*
Point A -> 110	1.34	4	0.335	110	90	[20 kN; 33 kN]	42	157.2	1524.91
110 -> 100	0.30	4	0.075	135	90	[11 kN; 76 kN]	77.8	65.19	632.39
100 -> 101	0.21	4	0.052	90	90	[17 kN; 0 kN]	23	13.36	129.62
101 -> 102	0.21	4	0.052	90	90	[17 kN; 0 kN]	23	13.36	129.62
102 -> 103	0.21	4	0.052	90	90	[17 kN; 0 kN]	23	13.36	129.62
103 -> 104	0.21	4	0.052	90	90	[17 kN; 0 kN]	23	13.36	129.62
104 -> 105	0.21	4	0.062	90	90	[17 kN; 0 kN]	23	13.36	129.62
105 -> 106	0.21	4	0.062	90	90	[17 kN; 0 kN]	23	13.36	129.62
106-> 96 (Point B)	0.30	4	0.075	135	90	[11 kN; 76 kN]	77.8	65.19	632.39
								SUM	3567.41

10.3.4.1 Scenario #1

In the navigation scenario depicted in Figure 10.12, the route determined by the algorithm to the first anchorage point is shown. This route was generated using environmental condition data specific to the area covered by the anchorage plan:

- Wind speed: 10 m/s
- True wind direction: 135°
- Sea current speed: 0.5 m/s
- True sea current direction: 90°

Using the transition weights provided in Figure 10.7, the route cost function can be calculated and is presented in Table 10.6. In this analysis, the cost associated with reaching the first waypoint has been excluded.

The route cost function for the first scenario, where the ship travels directly to the destination, is approximately 721, which is about 3% higher than the cost of the route determined by the algorithm. To compare the energy consumption of the direct route with the algorithmically determined route, an energy consumption analysis was conducted, and the results are summarised in Tables 10.7 and 10.8. To focus solely on the differences between the algorithmic route and the direct route to the destination, the first waypoint on the grid was designated as the starting point (A), while the green waypoint on the grid was set as the endpoint (B).

During the direct route to the destination, the energy consumption was 3121.34 kWh.

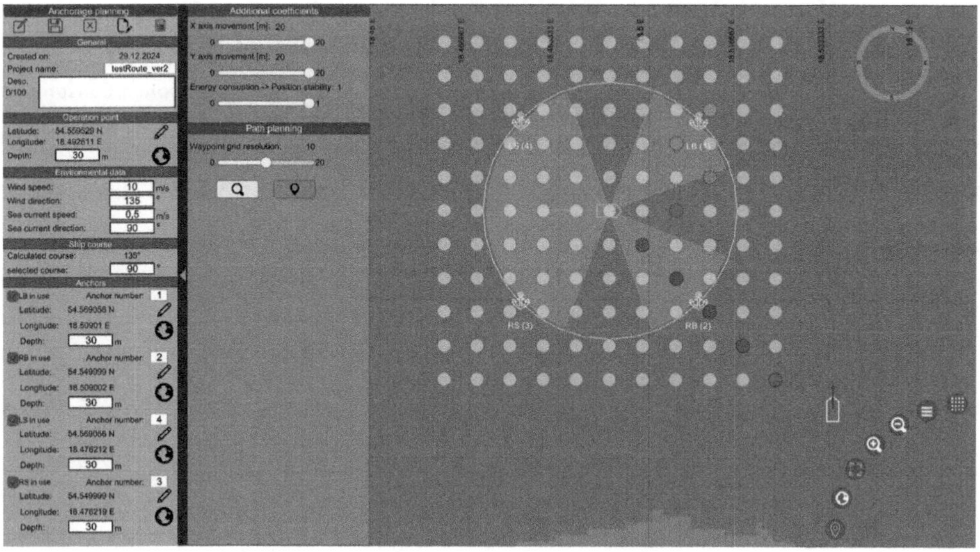

Figure 10.12 The ship passage route—true wind direction 135°.

Table 10.6 The value of the route cost function for navigation scenario #1

Waypoints' Edge	*d [m]*	*Course [°]*	*k [−]*	*d · κ*
10 -> 20	565	315	0.0611	34.52
20 -> 30	565	315	0.0611	34.52
30 -> 40	565	315	0.0611	34.52
40 -> 50	565	315	0.0611	34.52
50 -> 62	565	45	0.3115	175.99
62 -> 74	565	45	0.3115	175.99
74 -> 84	565	315	0.0611	34.52
84 -> 96	565	45	0.3115	175.99
			Sum	700.57

Table 10.7 Energy consumption for route without use of the algorithm

Phase	*S [NM]*	*V [kn]*	*t [h]*	*Course [°]*	*True Wind Angle [°]*	*Env. Force Vector [X-axis; Y-axis]*	*Sum of Thrust [kN]*	*Fuel Consumption [L]*	*Energy Consumption [kWh]*
Point A -> Point B	1.78	4	0.45	330	135	[25 kN; 47 kN]	64	321.78	3121.34

The energy consumption analysis revealed that the energy consumption for the route determined by the algorithm was 2951.42 kWh, which is about 6.5% lower than that of the direct route to the destination.

Table 10.8 Energy consumption for route with use of the algorithm

Phase	*S [NM]*	*V [kn]*	*t [h]*	*Course [°]*	*True Wind Angle [°]*	*Env. Force Vector [X-axis; Y-axis]*	*Sum of Thrust [kN]*	*Fuel Consumption [l]*	*Energy Consumption [kWh]*
Point A -> 20	0.30	4	0.075	315	135	[19 kN; 2 kN]	13.1	10.9	106.48
20 -> 30	0.30	4	0.075	315	135	[19 kN; 2 kN]	13.1	10.9	106.48
30 -> 40	0.30	4	0.075	315	135	[19 kN; 2kN]	13.1	10.9	106.48
40 -> 50	0.30	4	0.075	315	135	[19 kN; 2 kN]	13.1	10.9	106.48
50 -> 62	0.30	4	0.075	45	135	[0.8 kN; 99 kN]	99.2	83.1	806.34
62 -> 74	0.30	4	0.075	45	135	[0.8 kN; 99 kN]	99.2	83.1	806.34
74 -> 84	0.30	4	0.075	315	135	[19 kN; 2 kN]	13.1	10.9	106.48
84 -> Point B	0.30	4	0.075	45	135	[0.8 kN; 99 kN]	99.2	83.1	806.34
								SUM	2951.42

10.3.4.2 Scenario #2

The navigation scenario illustrated in Figure 10.13 presents the route calculated by the algorithm leading to the first anchorage point. This route was created using environmental data relevant to the area outlined in the anchorage plan:

- Wind speed: 10 m/s
- True wind direction: 45°
- Sea current speed: 0.5 m/s
- True sea current direction: 10°

Using the transition weights provided in Figure 10.7, the route cost function can be calculated and is presented in Table 10.9. In this analysis, the cost associated with reaching the first waypoint has been excluded once again.

The route cost function for the first scenario, where the ship travels directly to the destination, is approximately 1322.81, which is about 63% higher than the cost for the route calculated by the algorithm.

For the second scenario, an energy consumption analysis was performed under the same assumptions as those applied to the first scenario. The results of this analysis are presented in Tables 10.10 and 10.11.

During the direct route to the destination, the energy consumption was 2011.96 kWh.

In Scenario 2, the energy consumption for the route calculated by the algorithm was 1857.6 kWh, representing an 8% reduction compared to the energy consumption of the direct route.

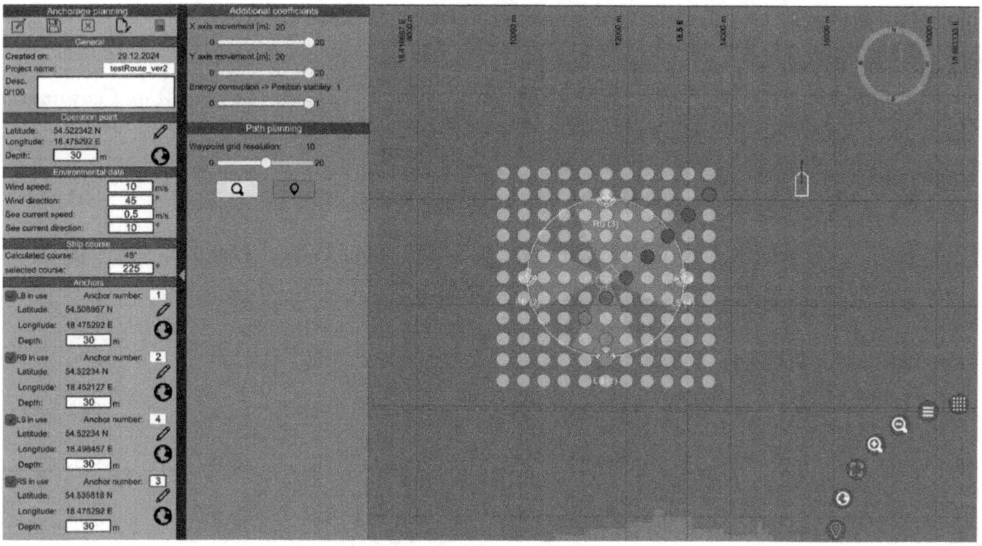

Figure 10.13 The ship passage route—true wind direction 45°.

Table 10.9 The value of the route cost function for navigation scenario #2

Waypoints' Edge	*d [m]*	*Course [°]*	*k [−]*	$d \cdot k$
109 -> 97	565	225	0.0611	34.52
97 -> 85	565	225	0.0611	34.52
85 -> 73	565	225	0.0611	34.52
73 -> 61	565	225	0.0611	34.52
61 -> 49	565	225	0.0611	34.52
49 -> 37	565	225	0.0611	34.52
37 -> 27	565	135	0.3115	175.99
27 -> 16	400	180	0.2439	97.56
			Sum	480.67

Table 10.10 Energy consumption for route without use of the algorithm

Phase	*S [NM]*	*V [kn]*	*t [h]*	*Course [°]*	*True Wind Angle [°]*	*Env. Force Vector [X-axis; Y-axis]*	*Sum of Thrust [kN]*	*Fuel Consumption [l]*	*Energy Consumption [kWh]*
Point A -> Point B	2.09	4	0.52	210	45	[19 kN; 22 kN]	35.7	207.41	2011.96

Table 10.11 Energy consumption for route with use of the algorithm

Phase	*S [NM]*	*V [kn]*	*t [h]*	*Course [°]*	*True Wind Angle [°]*	*Env. Force Vector [X-axis; Y-axis]*	*Sum of Thrust [kN]*	*Fuel Consumption [l]*	*Energy Consumption [kWh]*
Point A -> 97	0.3	4	0.075	225	45	[19 kN; 2 kN]	13.1	10.9	106.48
97 -> 85	0.3	4	0.075	225	45	[19 kN; 2 kN]	13.1	10.9	106.48
85 -> 73	0.3	4	0.075	225	45	[19 kN; 2 kN]	13.1	10.9	106.48
73 -> 61	0.3	4	0.075	225	45	[19 kN; 2 kN]	13.1	10.9	106.48
61 -> 49	0.3	4	0.075	225	45	[19 kN; 2 kN]	13.1	10.9	106.48
49 -> 37	0.3	4	0.075	225	45	[19 kN; 2 kN]	13.1	10.9	106.48
37 -> 27	0.3	4	0.075	135	45	[0.8 kN; 99 kN]	99	82.9	804.72
27 -> Point B	0.21	4	0.05	180	45	[14 kN; 76 kN]	76.4	42.6	414.01
								SUM	1857.6

10.4 Discussion

Section 3 introduces an algorithm designed to determine a local passage route by incorporating local hydrometeorological conditions and the ship's CP. The algorithm operates by dividing the navigational area into smaller grid squares (e.g., an area of 10.2 km² as illustrated in the publication), where environmental conditions are assumed to remain constant. The route is calculated within each square, and as the vessel moves from one area to the next, the route is recalculated based on the updated environmental conditions.

While the algorithm's structure is relatively straightforward, several potential limitations to its application were identified during the implementation process:

- **Availability of the Ship's Capability Plot:** As highlighted in the publication, any DP 2 or higher vessel is required to have a CP. However, if the route is being calculated for a DP 1 vessel or one lacking such a system, the CP would need to be generated independently.
- **Real-Time Environmental Monitoring:** If the route needs to be updated during operation, the vessel must have the capability to monitor wind speed in real time and either measure or estimate the speed of sea currents.
- **Impact of Programming Language on Performance:** The speed of route calculation is influenced by the programming language used for implementation. In the study, the algorithm was written in Python and integrated with code in C#. Both are high-level programming languages. If the algorithm were implemented in lower-level languages like C or C++, faster route generation times could be achieved. Currently, the computation times are as follows:

  For a 6×6 grid: 0.05 seconds
  For an 11×11 grid: 0.36 seconds

For a 16×16 grid: 0.47 seconds
For a 21×21 grid: 0.85 seconds

This publication also provides a comparison of the ship's passage routes in two scenarios: a direct route from the starting point to the endpoint and a route calculated using the algorithm. The analysis revealed that the algorithm-determined route reduced the environmental forces acting on the ship's hull by 3%, leading to a 5% decrease in energy consumption. In two additional scenarios, the algorithm demonstrated its effectiveness in reducing energy consumption. In the first case, the energy consumption was reduced by 6.5%, while in the second case, a reduction of 8% was achieved compared to the direct route.

The algorithm presented theoretically has no inherent limitations regarding the maximum length of the route it can determine, as long as real-time data on wind and ocean currents are available. However, during the studies conducted on route determination, it was not feasible to calculate routes spanning multiple areas with varying environmental conditions due to the inability to incorporate multiple hydrometeorological changes along the route. To demonstrate the algorithm's functionality, the example of a vessel navigating to an anchorage was chosen. Such vessels typically cover relatively short distances, allowing the entire operational area to be treated as having consistent environmental conditions.

The studies demonstrated that the presented algorithm could potentially be utilised in the future for the autonomous control of vessels required to anchor in accordance with a pre-planned anchorage layout.

10.5 Conclusions

The determination of ship passage routes is a critical aspect of maritime management, contributing to increased efficiency, enhanced safety, and the preservation of the marine environment. Properly planned passage routes help avoid hazardous areas such as shoals and reefs, significantly mitigating the risk of collisions, mechanical failures, and other incidents that endanger both the crew and the vessel. Effective routing systems and algorithms should account for dynamic sea conditions, continuously monitoring wind, currents, and waves, and updating the route as needed.

This study introduces a method for planning a ship's passage route using CPs in conjunction with an algorithm based on Dijkstra's methodology. By leveraging CPs to determine transition weights between waypoints, the method identifies routes that minimise the impact of wind on the ship's hull. However, this does not guarantee that the selected route satisfies other criteria, such as minimising travel time or being the shortest possible route. Nevertheless, for vessels lacking thrusters capable of enabling free movement along any axis, the method outlined in this publication offers a safer alternative.

The proposed route determination algorithm is versatile, capable of planning routes while the ship is in port and monitoring its progress during navigation. It recalibrates transition weights based on current environmental conditions, enabling dynamic route adjustments to ensure both efficiency and safety.

References

1. Przybylowski, A. Global trends shaping life quality in agglomerations with particular emphasis on mobility in seaport agglomerations. *TransNav* **2019**, *13*, 615–620. https://doi.org/10.12716/1001.13.03.18
2. Hanssen, G.L. The U.S. Navy hydrographic office ship routing program. *Navigation: Journal of the Institute of Navigation* **1958**, *6*(3), 157–165.

3. James, R.J. *Application of Wave Forecast to Marine Navigation.* Dissertation; New York University, 1957.
4. Pierson, W.J.; Neuman, G.; James, R.W. *Practical Methods for Observing and Forecasting Ocean Waves by Means of Wave Spectra and Statistics. (Reprinted 1971)*, 1955. U.S. Naval Hydrographic Office, Washington, DC.
5. Gao, M.; Shi, G.; Li, W.; Wang, Y.; Liu, D. an improved genetic algorithm for island route planning. *Procedia Engineering* **2017**, *174*, 433–441. https://doi.org/10.1016/j.proeng.2017.01.163
6. Yanchin, I.; Petrov, O. Parallel genetic algorithm for planning safe and optimal route for ship. *arXiv preprint arXiv:1905.05478,* 2019.
7. Kuhlemann, S.; Tierney, K. A genetic algorithm for finding realistic sea routes considering the weather. *Journal of Heuristics* **2020**, *26*, 801–825. https://doi.org/10.1007/s10732-020-09449-7
8. Zhao, W.; Wang, Y.; Zhang, Z.; Wang, H. Multicriteria ship route planning method based on improved particle swarm optimization–genetic algorithm. *JMSE* **2021**, *9*, 357. https://doi.org/10.3390/jmse9040357
9. He, Y.; Li, Z.; Mou, J.; Hu, W.; Li, L.; Wang, B. Collision-avoidance path planning for multi-ship encounters considering ship manoeuvrability and COLREGs. *Transportation Safety and Environment* **2021**. https://doi.org/10.1093/tse/tdab004
10. Wang, L.; Zhang, Z.; Zhu, Q.; Ma, S. Ship route planning based on double-cycling genetic algorithm considering ship maneuverability constraint. *IEEE Access* **2020**, *8*, 190746–190759. https://doi.org/10.1109/ACCESS.2020.3031739
11. Wu, G.; Li, Y.; Jiang, C.; Wang, C.; Guo, J.; Cheng, R. Multi-vessels collision avoidance strategy for autonomous surface vehicles based on genetic algorithm in congested port environment. *Brod* **2022**, *73*, 69 91. https://doi.org/10.21278/brod73305
12. Wu, Y.; Wang, T.; Liu, S. a review of path planning methods for marine autonomous surface vehicles. *JMSE* **2024**, *12*, 833. https://doi.org/10.3390/jmse12050833
13. Jang, D.; Kim, J. Development of ship route-planning algorithm based on Rapidly Exploring Random Tree (RRT*) using designated space. *JMSE* **2022**, *10*, 1800. https://doi.org/10.3390/jmse10121800
14. Pennino, S.; Gaglione, S.; Innac, A.; Piscopo, V.; Scamardella, A. Development of a new ship adaptive weather routing model based on seakeeping analysis and optimization. *JMSE* **2020**, *8*, 270. https://doi.org/10.3390/jmse8040270
15. Walther, L.; Rizvanolli, A.; Wendebourg, M.; Jahn, C. Modeling and optimization algorithms in ship weather routing. *International Journal of e-Navigation and Maritime Economy* **2016**, *4*, 31–45. https://doi.org/10.1016/j.enavi.2016.06.004
16. Zhang, Y.; Wen, Y.; Tu, H. A method for ship route planning fusing the ant colony algorithm and the A* search algorithm. *IEEE Access* **2023**, *11*, 15109–15118. https://doi.org/10.1109/ACCESS.2023.3243810
17. Lee, J.-S.; Kim, T.-H.; Park, Y.-G. Maritime transport network in Korea: spatial temporal density and path planning. *JMSE* **2023**, *11*, 2364. https://doi.org/10.3390/jmse11122364
18. Silveira, P.; Teixeira, Â.P.; Guedes Soares, C. AIS based shipping routes using the Dijkstra algorithm. *TransNav* **2019**, *13*, 565–571. https://doi.org/10.12716/1001.13.03.11
19. Li, Y.; Cui, J.; Zhang, X.; Yang, X. A ship route planning method under the sailing time constraint. *JMSE* **2023**, *11*, 1242. https://doi.org/10.3390/jmse11061242
20. Chen, G.; Wu, T.; Zhou, Z. Research on ship meteorological route based on a-star algorithm. *Mathematical Problems in Engineering* **2021**, *2021*, 1–8. https://doi.org/10.1155/2021/9989731
21. Grifoll, M.; Borén, C.; Castells-Sanabra, M. A comprehensive ship weather routing system using CMEMS products and A* algorithm. *Ocean Engineering* **2022**, *255*, 111427. https://doi.org/10.1016/j.oceaneng.2022.111427
22. Liu, C.; Mao, Q.; Chu, X.; Xie, S. An Improved A-Star Algorithm Considering Water Current, Traffic Separation and Berthing for Vessel Path Planning. *Applied Sciences* **2019**, *9*, 1057, doi:10.3390/app9061057

23. Wang, Y.; Qian, L.; Hong, M.; Luo, Y.; Li, D. Multi-objective route planning model for ocean-going ships based on bidirectional a-star algorithm considering meteorological risk and IMO guidelines. *Applied Sciences* **2024**, *14*, 8029. https://doi.org/10.3390/app14178029
24. Tsou, M.-C.; Cheng, H.-C. An ant colony algorithm for efficient ship routing. *Polish Maritime Research* **2013**, *20*, 28–38. https://doi.org/10.2478/pomr-2013-0032
25. Zhang, M.; Ren, H.; Zhou, Y. Research on global ship path planning method based on improved ant colony algorithm. *The IEEE Open Journal of Intelligent Transportation Systems* **2023**, *4*, 143–152. https://doi.org/10.1109/OJITS.2023.3247377
26. Akka, K.; Khaber, F. Mobile robot path planning using an improved ant colony optimization. *International Journal of Advanced Robotic Systems* **2018**, *15*. https://doi.org/10.1177/1729881418774673
27. Zhang, D.; You, X.; Liu, S.; Pan, H. Dynamic multi-role adaptive collaborative ant colony optimization for robot path planning. *IEEE Access* **2020**, *8*, 129958–129974. https://doi.org/10.1109/ACCESS.2020.3009399
28. Morin, M.; Abi-Zeid, I.; Quimper, C.-G. Ant colony optimization for path planning in search and rescue operations. *European Journal of Operational Research* **2023**, *305*, 53–63. https://doi.org/10.1016/j.ejor.2022.06.019
29. Zhang, H.; Sun, J.; Yang, B.; Shi, Y.; Li, Z. Optimal search and rescue route design using an improved ant colony optimization. *ITC* **2020**, *49*, 438–447. https://doi.org/10.5755/j01.itc.49.3.25295
30. Tsou, M.-C.; Hsueh, C.-K. The study of ship collision avoidance route planning by ant colony algorithm. *Journal of Marine Science and Technology* **2020**, *18*. https://doi.org/10.51400/2709-6998.1929
31. Shen, Y.; Wang, F.; Zhao, P.; Tong, X.; Huang, J.; Chen, K.; Zhang, H. Ship Route Planning Based on Particle Swarm Optimization. In *2019 34rd Youth Academic Annual Conference of Chinese Association of Automation (YAC)*. 2019 34rd Youth Academic Annual Conference of Chinese Association of Automation (YAC), Jinzhou, China, 06–08 Jun. 2019; IEEE, 2019; pp 211–215.
32. Yujie, L.; Yu, P.; Yixin, S.; Huajun, Z.; Danhong, Z.; Yong, S. Ship Path Planning Based on Improved Particle Swarm Optimization. In *2018 Chinese Automation Congress (CAC)*. 2018 Chinese Automation Congress (CAC), Xi'an, China, 30 Nov.–02 Dec. 2018; IEEE, 2018; pp 226–230.
33. Li, B.; Chiong, R.; Gong, L. Search-evasion path planning for submarines using the Artificial Bee Colony algorithm. In *2014 IEEE Congress on Evolutionary Computation (CEC)*. 2014 IEEE Congress on Evolutionary Computation (CEC), Beijing, China, 06–11 Jul. 2014; IEEE, 2014; pp 528–535.
34. Abramowicz-Gerigk, T.; Burciu, Z.; Jaworski, T.; Nowicki, J. The large-scale physical model tests of the passing ship effect on a ship moored at the solid-type berth. *Sensors* **2022**, *22*. https://doi.org/10.3390/s22030868
35. Fossen, T.I. *Handbook of Marine Craft Hydrodynamics and Motion Control,* Wiley, Chichester, 2011.
36. Stewart, R. Ocean-Wave Spectra. https://wikiwaves.org/Ocean Wave_Spectra
37. DNV. Main website. https://www.dnv.com/index.htm
38. Johansen, T.A.; Fuglseth, T.P.; Tøndel, P.; Fossen, T.I. Optimal constrained control allocation in marine surface vessels with rudders. *Control Engineering Practice* **2008**, *16*, 457–464. https://doi.org/10.1016/j.conengprac.2007.01.012

Chapter 11

Toward reliable PNT data for autonomous shipping

Natalia Ziółkowska and Joanna Szłapczyńska

11.1 Introduction

Reliable positioning, navigation, and timing (PNT) data are particularly relevant today in the context of the rapidly developing field of autonomous maritime transport. Despite their pivotal role, an existing PNT system often faces challenges such as signal degradation, interference, and susceptibility to cyberattacks [1]. These problems require innovative solutions to improve the accuracy and robustness of PNT data.

The introduction of Maritime Autonomous Surface Ships (MASS) notion by the International Maritime Organization (IMO) in 2018 paved the way toward advanced and robust marine transportation. When one envisions autonomous shipping accompanied by alternative fuels and zero-emission propulsion systems, the prospective advancements in the maritime industry [2,3] might seem dazzling. However, evolution toward this goal raises key concerns, particularly in the domains of safety, security, and environmental protection.

Reliable PNT data is critical for ensuring the safe operation of MASS, thereby enabling autonomous systems to make accurate real-time decisions. In this regard, modern propulsion systems, such as hybrid engines that combine conventional and renewable energy sources, solar-powered technologies, and next-generation sails that utilize rigid wings and AI-driven optimization [4], have emerged as viable solutions for sustainable shipping [5,6]. Improving ship safety and robustness is of great importance, especially in light of the growing interest in nuclear propulsion [7], which is currently employed in military vessels and icebreakers [8] and holds potential for expansion into commercial shipping. Such developments align with the industry's green propulsion initiatives aimed at reducing greenhouse gas emissions [9]. Equally crucial in the context of safe transportation is the reliable operation of propulsion systems, where artificial intelligence (AI) is increasingly utilized for anomaly detection and mitigation [10].

MASS represents a significant advancement in maritime technology, promising to enhance operational efficiency, reduce costs, and, to some extent, improve safety. The IMO has identified four degrees of autonomy [11], ranging from partial automation to fully autonomous ships:

1 Ship with automated processes and decision support.
2 Remotely controlled ship with crew on board.
3 Remotely controlled ship without crew on board.
4 Fully autonomous ship.

The first degree indicates the presence of automated processes and decision support systems on a vessel. The crew remains on board, engaged in monitoring and controlling onboard systems.

DOI: 10.1201/9781003686088-11

While some operations may be automated, the crew is prepared to perform necessary tasks if required. The second degree of autonomy comprises a remotely operated vessel with a limited onboard crew. Operations are controlled from another location, but the crew remains on board to intervene if necessary. In the third degree, the ship is also operated remotely, but in this case, there is no crew on board. A ship with the fourth degree of autonomy is one that is fully autonomous, meaning that it is capable of independently executing decisions and operational actions through the use of automated systems. Furthermore, the vessel is prepared to autonomously respond to an emergency scenario.

It is important to highlight that reliable PNT data is indispensable across all levels of autonomy. Additionally, as the level of autonomy increases, the reliance on accurate and robust PNT data grows significantly. This is due to the fact that the autonomy of ships introduces greater uncertainties and ambiguities [12]. These challenges primarily stem from the following factors: (1) the presence of incomplete or inaccurate data, (2) the limited confidence in system judgments, and (3) the reliance on underdeveloped models for assessing and mitigating vulnerabilities.

The integration of advanced navigation and positioning systems in MASS plays a vital role in ensuring their safe and effective operation. As highlighted in [13], inaccuracies in data inputs, particularly those derived from computer vision and real-time sensors, pose significant risks to navigation and safety.

This paper seeks to address these challenges by focusing on improving the reliability and accuracy of PNT data. Additionally, it aims to enable effective handling of uncertainties in the marine environment through the application of AI methodologies. The proposed conceptual framework centers on AI-driven approaches to mitigate errors and enhance the signal integrity of PNT data. By demonstrating the impact of these solutions on the safety and operational efficiency of autonomous vessels, this study aims to contribute to the advancement of MASS technologies.

The rest of the paper is organized as follows: Works related to the research on PNT are briefly discussed in "Related works", which is followed by a discussion on PNT data application to autonomous shipping and issues related to the process presented in "Autonomous vessel issues related to PNT data". The proposed AI-based solution is described in "Proposed ML-based solution for improved PNT data". Finally, conclusions are given in "Conclusions".

11.2 Related works

Recent studies have emphasized the critical role of PNT data in supporting maritime operations [14,15], with the U.S. Department of Transportation underscoring the necessity of robust and resilient PNT systems to ensure reliable data [16]. The PNT system comprises three main components. The shipboard component encompasses devices and sensors that facilitate the transmission of information regarding the vessel's status and operations. The onshore component includes the support infrastructure for navigation on land. Lastly, the communication link facilitates data exchange between these two components.

As illustrated in Figure 11.1, the PNT system encompasses a wide range of sensors and data sources that provide critical information to the vessel. These include Global Navigation Satellite Systems (GNSS) such as GPS, Galileo, GLONASS, and BeiDou, which serve as primary sources of positioning data. Moreover, there are gyroscopes, accelerometers, magnetometers, radar, and Inertial Navigation Systems (INS). Additionally, inputs from sonar and visual sensors, including cameras for optical navigation, contribute to the data stream. While these devices collectively enhance the system's accuracy and reliability, it is important to note that some information may be erroneous due to factors such as signal interference, sensor malfunctions, or deliberate spoofing.

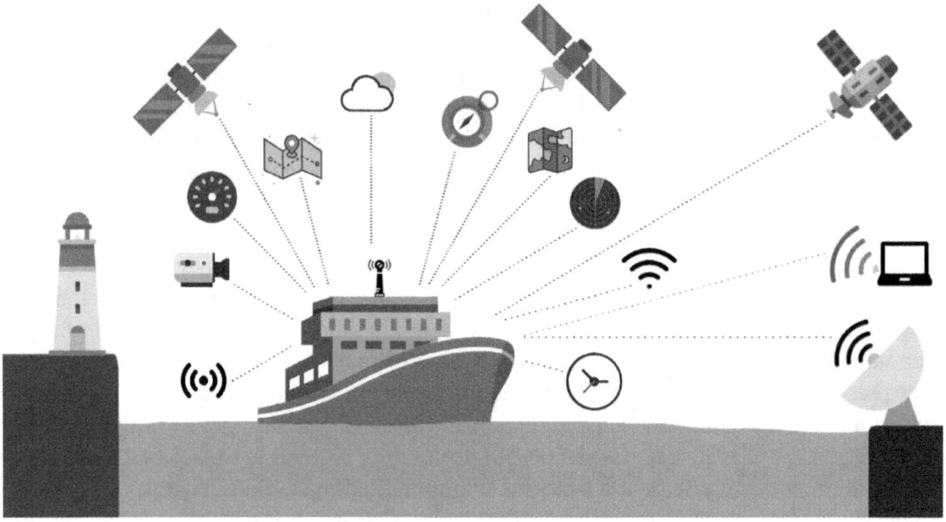

Figure 11.1 Architecture of the PNT System: Integration of Data Sources.

It is noteworthy that a PNT system, even when operating in a complex environment, should maintain consistency, accessibility, and reliability [17]. This is particularly relevant in contexts where the complexity of factors affecting PNT data is increasing, necessitating robust mechanisms to ensure data reliability. Core functions of a resilient PNT framework should include error prevention, responding to anomalies, and recovering from problems (if required) [18]. Consequently, it is advisable to leverage devices from multiple sources and propose an intelligent fusion of PNT data [19]. This approach not only enhances the system's resilience to failures in individual components but also ensures the delivery of higher-quality information – a crucial aspect, particularly relevant for autonomous vessels. Reliable PNT data is essential for various maritime applications, including ship weather routing [20] and collision avoidance [21,22].

A review of the literature indicates two main approaches to improving PNT data quality: traditional methods and modern AI-based solutions. With regard to traditional methods of enhancing PNT data, filters for real-time data prediction and correction are frequently employed. The Kalman Filter and its modifications are prominent examples in this category [23]. Alternatively, Dempster-Shafer theory (evidence theory) has demonstrated effectiveness in scenarios characterized by uncertainty and incomplete data [24,25]. In addition, trajectory interpolation methods have been employed to address the issue of missing data. Traditional methods of improving PNT data are effective in stable environments, however their effectiveness is reduced in dynamic and non-deterministic conditions. Furthermore, traditional approaches demonstrate a lack of adaptability to complex scenarios, such as sudden course changes or GNSS signal interference [26].

AI techniques are increasingly employed to overcome the limitations of conventional navigation systems. The utilization of AI-based solutions, including neural networks, along with other algorithms, facilitates a more adaptive approach to data processing.

Recurrent neural networks (RNNs) are a variant of neural networks specifically designed to process sequential data by maintaining an internal state that captures temporal dependencies across time steps. This ability makes RNNs well-suited for modeling time series and sequential patterns, such as marine data fusion [27].

Convolutional neural networks (CNNs) represent a type of artificial neural network that was originally developed for the analysis of images. However, they have since become a widely utilized tool for data with spatial structure or local dependencies, including images, maps, and signals. The key feature of a CNN is the use of convolutional layers, which automatically learn to extract important features from the input data. These tools have been demonstrated to be effective in the identification of anomalies in spatial data [28].

Support Vector Machines (SVMs) are a class of supervised learning models used for classification and regression tasks. In maritime navigation, SVMs have demonstrated effectiveness in sensor fusion [29] by enabling the integration of heterogeneous sensor data for improved positioning accuracy.

These techniques enable the prediction of missing data based on historical trajectory analysis. However, they are not without limitations. CNNs are designed for Euclidean data, like images, where data is structured on a uniform grid. This type of neural network is particularly sensitive to spatial distortions and may lack the capacity to generalize when applied to environments that diverge from their training data [30,31]. Similarly, while RNNs demonstrate proficiency in processing sequential data, they encounter challenges such as vanishing gradient problems when confronted with extensive time-series predictions [32]. Furthermore, despite their demonstrated capabilities, the AI-based approaches previously mentioned are prone to limitations such as computational demands, scalability issues, and sensitivity to data quality. These factors must be addressed to ensure reliable operation in maritime environments.

Although research on improving the quality of PNT data is ongoing, the potential of integrating AI techniques to achieve comprehensive reliability and accuracy remains underexplored. Further investigation into AI-driven data fusion within the context of maritime transportation is necessary.

11.3 Autonomous vessel issues related to PNT data

The reliability of PNT data is a critical issue in maritime navigation, especially given the increasing reliance on technology and automation in the development of autonomous vessels [33,34]. Imprecise or unreliable PNT data can lead to threats to the safety of people, cargo, and the environment. Navigation system failures, sensory issues, and unpredictable vessel behavior are among the top safety concerns [35]. Sensory issues, such as the presence of delayed data from a single sensor, in conjunction with latency and processing errors, have the potential to disrupt the integration of information from multiple sensors. This disruption can result in inconsistencies in the vessel's perception of its environment.

A primary problem that may be faced by autonomous vessels is their susceptibility to temporary disruptions in GNSS signals. The signal may be completely lost, or its availability may be intermittent. Furthermore, the frequency of updates may be reduced. In such scenarios, the continuity of positional data may be at risk. These interferences may be attributed to various factors, including environmental disturbances, malfunctions in the device itself, or deliberate actions such as jamming or spoofing. Such disruptions can lead to significant navigation errors and a reduction in the vessel's ability to respond to the dynamic maritime environment. The ability of a ship to maintain reliable navigation depends on the extent to which data can be extrapolated using predictive algorithms or onboard inertial systems.

Contemporary systems have the capacity to maintain satisfactory accuracy for brief intervals; however, errors typically accumulate over time. In dynamic maritime scenarios, this time frame is often limited before the extrapolated data no longer meets the accuracy requirements for safe navigation.

Table 11.1 Selected metrics for assessing the reliability of GNSS signals in marine navigation

Metric name	*Shortened name*	*Description*	*Purpose*
Signal-to-noise ratio	SNR	Ratio of received signal power to background noise.	Indicates signal quality; higher SNR means better signal clarity and reliability.
Carrier-to-noise density ratio	C/N_0	Ratio of carrier power to noise power density.	A more precise indicator of signal quality at the GNSS receiver antenna input.
Position dilution of precision	PDOP	Geometric factor indicating the effect of satellite geometry on 3D position accuracy.	Helps assess the potential accuracy of 3D position fixes.
Horizontal dilution of precision	HDOP	Geometric factor for horizontal positioning accuracy (latitude and longitude).	Indicates the expected accuracy of horizontal position fixes.
Vertical dilution of precision	VDOP	Geometric factor for vertical positioning accuracy (altitude).	Indicates the expected accuracy of vertical position fixes.
Time dilution of precision	TDOP	Geometric factor related to timing accuracy.	Important for systems relying on precise GNSS timing (e.g., synchronization).
Geometric dilution of precision	GDOP	Combined factor including position and time dilution of precision.	Provides an overall view of satellite geometry impact on position and time accuracy.
Number of satellites in view	N_{sat}	Number of GNSS satellites that are currently visible to the receiver.	More satellites generally improve positioning accuracy and reliability.
Protection level	PL	Bound on data error with a specified confidence level.	Used in safety-critical applications to ensure the data meets integrity requirements.
Integrity risk	–	Probability that the system outputs data exceeding the alert limit without warning.	Critical for high-reliability applications; ensures the system detects faults or degradation.
Availability	–	Percentage of time the GNSS service provides usable data with the required performance.	Assesses system reliability over time.
Continuity	–	Probability that the system will provide data without interruption during an operational phase.	Important for phases requiring uninterrupted navigation even under dynamic conditions.
Multipath error estimation	–	Estimation of errors caused by reflected GNSS signals interfering with the direct signal.	Important in harbors or near large structures such as bridges, where signal reflections can significantly degrade positioning accuracy.

Another critical challenge is the absence of confidence indicators for PNT data reliability. Current PNT systems often do not provide sufficient metrics to assess the reliability of their data, resulting in a lack of clarity for operators and autonomous systems regarding the potential inaccuracies or vulnerabilities inherent in navigation inputs. Trusting data with an unknown degree of reliability can lead to erroneous navigation decisions. Errors or delays in PNT data processing can cause operators to make decisions based on outdated or incorrect information.

In Table 11.1 selected metrics used to assess the reliability of GNSS signals in marine navigation are presented. These metrics provide information on signal quality, accuracy, integrity, and potential error sources, which are essential for ensuring safe and accurate vessel navigation.

Selected metrics, especially signal-to-noise ratio (SNR) [36] and dilution of precision (DOP) [37], play an important role in assessing data quality. SNR measures the ratio of desired signal strength to background noise, providing an indication of signal clarity and reliability. Higher SNR values correspond to better signal quality, which ensures more accurate positioning data. Meanwhile, DOP evaluates the impact of satellite geometry on the accuracy of positioning and timing data in GNSS. A lower DOP value reflects a favorable satellite configuration resulting in improved accuracy, while higher DOP values indicate degraded accuracy due to less than optimal satellite alignment.

Advanced metrics should encompass accuracy, integrity, continuity, and availability to provide a comprehensive assessment of PNT data reliability and effectiveness [38]. One such metric is integrity risk, which quantifies the probability of the system delivering inaccurate data without timely user warning. This metric measures the frequency of instances where PNT data exceed acceptable error thresholds without triggering an integrity alert, thus emphasizing system safety and trustworthiness. Similarly, continuity represents the likelihood of disruptions in PNT data delivery that could prevent consistent usability during critical operations. This metric evaluates the system's ability to ensure data delivery remains within acceptable limits even under dynamic conditions. Metrics such as integrity risk, availability, and continuity are not typically provided directly by GNSS receivers. Instead, they are defined in system specifications and performance standards and are usually assessed through long-term system performance analyses rather than real-time receiver output.

11.4 Proposed ML-based solution for improved PNT data

The need for scalable and adaptable PNT solutions that can address diverse maritime scenarios is increasingly evident. The identified gaps underscore the necessity of the framework proposed in this study to address these limitations and establish a solid foundation for future research in the domain of MASS positioning.

To effectively confront these challenges, innovative solutions must be implemented to mitigate environmental impacts, incorporate confidence measures for PNT data, and enhance system resilience against signal interruptions and interference. The integration of advanced machine learning (ML) algorithms and sensor fusion techniques offers considerable potential for addressing these issues, thereby enabling the development of more robust and reliable autonomous maritime navigation systems.

This paper proposes a novel approach that combines graph neural networks (GNN) and generative adversarial networks (GANs). By integrating these technologies, the framework aims to create a robust system capable of improving the reliability and accuracy of navigation data. The review [39] highlights the growing importance of advanced neural network architectures, including GNN and GAN, for detecting and mitigating anomalies in time-series data across diverse application domains. GNN demonstrate proficiency in relational data analysis, while GAN offer a robust framework for trajectory reconstruction. These capabilities support their selection for applications in PNT, particularly in the maritime environment where data complexity and real-time requirements are challenging. In this framework, GANs are employed to generate high-quality training data and to perform data and trajectory reconstruction, thereby enhancing the training process. The GNN model is then trained on this data to improve the quality of navigation data and to effectively handle uncertainty when applied to new, unseen navigation data. In this architecture, the GNN serves as the core model for inference, while the GAN facilitates robust preparation of the training dataset.

11.5 Generative adversarial networks

GANs consist of two competing networks: a generator and a discriminator. The learning process occurs through competition between these two neural networks. As illustrated in Figure 11.2, the generator creates new data based on provided patterns, while the discriminator evaluates the quality of the data generated. In the context of maritime navigation, GANs have significant potential, particularly for reconstructing incomplete or missing ship data even under extreme conditions. By generating plausible synthetic data that aligns with existing patterns, GAN can mitigate disruptions, ensuring the continuity and reliability of navigation systems. This is especially critical in scenarios where environmental conditions or equipment failures compromise the availability of accurate position and timing data. This capability can enhance the continuity and reliability of navigation information, even in the absence of some PNT information.

The application of GAN models in the domain of maritime transportation is beginning to emerge. One notable instance is the use of GAN-AI models in trajectory prediction [40], where a model designed to predict trajectories between multiple vessels exhibited superior accuracy in scenarios involving complex ship interactions, thereby enhancing prediction performance by up to 72% compared to traditional models. Another example is the GAN-based Automatic Identification System (AIS) data repair model (TLGAN) [41], which was developed to address the challenges posed by missing and anomalous data in marine navigation systems. This model demonstrates the potential of GANs to enhance data reliability and safety in navigation.

The integration of GAN into maritime positioning systems presents a transformative opportunity to address critical challenges such as data absence, anomalies, and the need for data

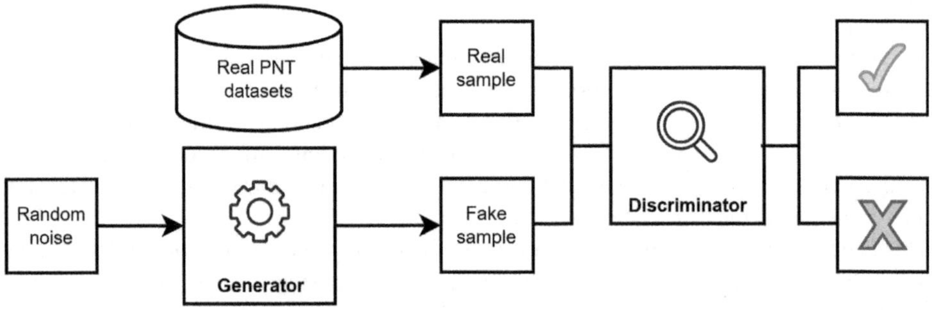

Figure 11.2 Schematic diagram of the proposed generative adversarial network (GAN) architecture for PNT data generation and reconstruction.

continuity in diverse and uncertain marine environments. This is especially important in scenarios where environmental conditions or equipment failures compromise the availability of accurate position and timing data.

11.6 Graph neural networks

Graph neural networks (GNN) offer unique advantages for maritime navigation, particularly due to their design, which is tailored for analyzing data represented as graphs. Unlike traditional neural networks, GNNs are specifically designed to handle graph-based data, which naturally represents maritime navigation scenarios such as ship trajectories, communication networks, or spatial relationships between vessels and environmental components. As shown in Figure 11.3, these networks leverage the properties of graphs, such as local links and global relationships, enabling efficient analysis of spatial and temporal data. GNNs demonstrate considerable potential in maritime navigation due to their ability to capture spatial-temporal relationships, positioning GNNs as a promising tool for advancing PNT systems.

One potential application of GNN would be the analysis of ship trajectories as a sequence of related points in space and time. For example, ship trajectories can be viewed as sequences of interconnected spatiotemporal points, where the relationship between neighboring elements provides critical insights into positional accuracy. By considering the relationship between neighboring points of the trajectories, the objective is to improve position accuracy. Another potential application of this technology involves real-time detection, such as identifying incorrect or inconsistent data, ensuring the reliability of navigation inputs.

Real-time anomaly detection is another promising application of GNNs in maritime navigation. These networks can identify incorrect or inconsistent data in navigation inputs by modeling the dependencies between sensors and environmental factors. This ensures the reliability and safety of maritime operations, particularly in scenarios where accurate and timely decision-making is critical. The article [42] explores a method for monitoring and detecting anomalous behavior using GNN, specifically Graph Attention Network (GAT), to improve the real-time detection and accuracy of ship anomaly identification. Another investigation [43] presents an approach to recognizing ship behavior patterns using graph deep learning techniques combined with network modeling. These advancements highlight the potential of GNN to enhance the precision, safety, and efficiency of the PNT system at sea.

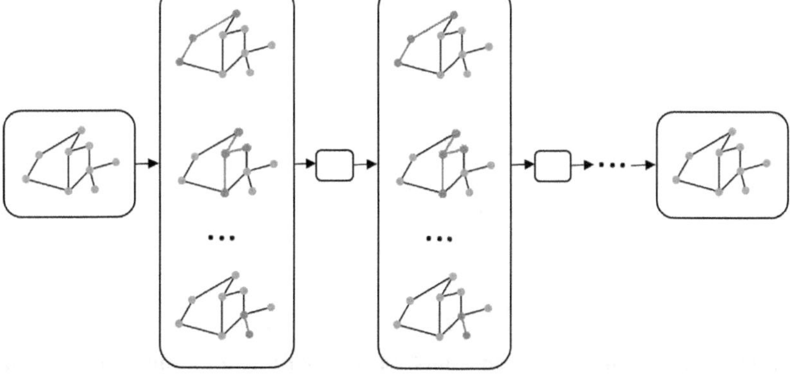

Figure 11.3 Schematic representation of the graph neural network (GNN) architecture.

11.7 Integrating GAN and GNN for reliable PNT data

To address the challenges of noisy, incomplete, and uncertain PNT data, this work proposes a novel framework that integrates GANs and GNN.

In the first stage, a GAN-based model is employed to reconstruct and augment historical PNT data. The generator learns to synthesize realistic PNT data patterns based on historical examples, while the discriminator ensures that the generated data is indistinguishable from authentic observations. This process not only fills in missing data but also enhances the diversity and completeness of the training set, which is critical for effective learning in subsequent stages.

In the second stage, the GNN is trained on both real and GAN-augmented data. The GNN architecture leverages spatiotemporal relationships within the data by representing PNT observations as a graph, where nodes correspond to measurement points and edges capture temporal or sensor-derived correlations. Through iterative message passing and feature aggregation, the GNN learns to enhance the accuracy, reliability, and robustness of PNT data. Once trained, the GNN becomes the core model capable of processing new incoming GNSS/INS data and improving output position.

As shown in the flowchart (Figure 11.4), the proposed solution to enhance the accuracy of navigation data consists of three key components: assessment of GNSS signal reliability, positioning with GNSS, and positioning without GNSS.

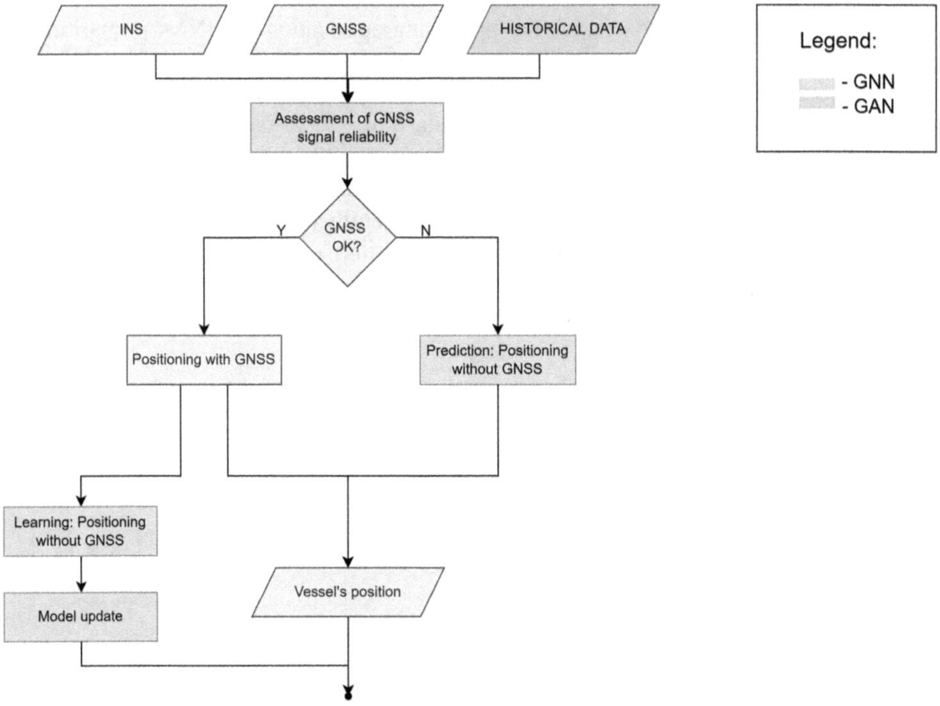

Figure 11.4 Diagram of the proposed GNN-based solution for improved PNT data. (GAN, Generative Adversarial Networks; GNN, Graph Neural Networks; GNSS, Global Navigation Satellite Systems; INS, Inertial Navigation Systems.)

First, before performing any calculations or navigational decisions, it is crucial to assess the reliability of GNSS signals. Any anomalies or deviations from expected signal characteristics may indicate interference or other sources of signal degradation. This assessment is conducted using GNN that compares signal characteristics with expected patterns. Identified anomalies serve as indicators of potential interference or degradation. Based on this verification, the system determines whether to continue using GNSS signals in further operations.

In the presence of reliable GNSS data, the system integrates these signals with other sources. Conversely, in the absence of reliable data, the system switches to prediction mode, based on PNT data sources, such as INS and historical data, provided that the available data supports accurate predictions. In cases where data is limited, the system estimates the expected positioning error, enabling navigation by making informed decisions under increased uncertainty.

The proposed GNN-based framework processes raw PNT data obtained from GNSS and INS sensors. The raw measurements are first represented as a spatiotemporal graph, where nodes correspond to individual PNT observations, and edges capture temporal or spatial dependencies. Through iterative message passing across the graph layers, the GNN learns complex relationships between measurements, enabling noise reduction, uncertainty management, and overall improvement of positioning accuracy.

This integrated framework aims to enhance the state-of-the-art in autonomous maritime navigation by addressing existing limitations and providing a scalable, resilient solution adaptable to diverse scenarios. With this approach, the PNT system not only enhances the safety of autonomous vessels, but also improves their operational efficiency, enabling accurate and reliable operation even under the most demanding conditions. This is a key step toward fully autonomous maritime systems.

11.8 Conclusions

To sum up, the proposed AI-based PNT solution is expected to significantly enhance the safety and efficiency of autonomous vessels. Firstly, the key benefit is the increased resilience to interference, achieved through the detection and mitigation of anomalies and the pursuit of reliable positioning continuity. Another aspect is the improved accuracy of positioning, facilitated by modern AI methods that analyze spatiotemporal relationships to increase the accuracy of PNT data. Additionally, the ability to manage uncertainty, thanks to the intelligent fusion of data from multiple sources, is another benefit. Furthermore, accurate information derived from PNT data sources is vital for making informed operational decisions, thereby minimizing the risk of erroneous navigational decisions.

Future directions toward reliable PNT data for autonomous vessels include the implementation and testing of the proposed solution in dynamic marine environments across diverse scenarios, including cases of measurement absence or inaccuracy, and basic protection against signal attacks. By addressing these areas, the maritime industry would ensure that future PNT systems are not only robust and accurate but also adaptable to the evolving demands of autonomous navigation.

Key areas for further exploration include verifying the real-time positioning of GNN on PNT data. Additionally, comparing the efficacy of GNN with other data fusion techniques – both traditional and AI-based – will provide valuable insights. The application of GAN for predicting missing trajectory points is also critical. Initial efforts should focus on using GAN to predict absent spatiotemporal coordinates based on historical data, progressing toward real-time prediction during instances of data loss to estimate the vessel's position under challenging conditions.

References

1. Pawelski, Jan. 2023. "Cyber Threats for Present and Future Commercial Shipping." *TransNav, International Journal on Marine Navigation and Safety Od Sea Transportation* 17(2): 261–267. https://doi.org/10.12716/1001.17.02.01
2. Kim, Mingyu, Tae-Hwan Joung, Byongug Jeong, and Han-Seon Park. 2020. "Autonomous Shipping and Its Impact on Regulations, Technologies, and Industries." *Journal of International Maritime Safety, Environmental Affairs, and Shipping* 4(2): 17–25. https://doi.org/10.1080/25725084.2020.1779427
3. Eide, Egil, Morten Breivik, Edmund F. Brekke, Bjørn-Olav H. Eriksen, Erik. Wilthil, Øystein K. Helgesen, Emil H. Thyri, Erik Veitch, Ole Andreas Alsos, and Tor Arne Johansen. 2025. "The Autonomous Urban Passenger Ferry milliAmpere2: Design and Testing." *Journal of Offshore Mechanics and Arctic Engineering* 147(031409). https://doi.org/10.1115/1.4067370
4. Inal, Omer Berkehan, Jean-Frédéric Charpentier, and Cengiz Deniz. 2022. "Hybrid Power and Propulsion Systems for Ships: Current Status and Future Challenges." *Renewable and Sustainable Energy Reviews* 156(March): 111965. https://doi.org/10.1016/j.rser.2021.111965
5. Editorial, RINA. 2024. "Wind Propulsion 2024: Advancements in Sails." https://rina.org.uk/publications/the-naval-architect/wind-propulsion-2024-advancements-in-sails/.
6. Nyanya, Mphatso N., Huy B. Vu, Alessandro Schönborn, and Aykut I. Ölçer. 2021. "Wind and Solar Assisted Ship Propulsion Optimisation and Its Application to a Bulk Carrier." *Sustainable Energy Technologies and Assessments* 47(October): 101397. https://doi.org/10.1016/j.seta.2021.101397
7. Głuch, Jerzy, Tomasz Kodlewicz, Marta Drosińska-Komor, Natalia Ziółkowska, Łukasz Breńkacz, and Paweł Ziółkowski. 2024. "Thermodynamic Efficiency of an Advanced 4th Generation VHTR Propulsion Engine for Large Container Ships." *Polish Maritime Research* 31(4): 76–88. https://doi.org/10.2478/pomr-2024-0052
8. Freire, Luciano Ondir, and Delvonei Alves de Andrade. 2015. "Historic Survey on Nuclear Merchant Ships." *Nuclear Engineering and Design* 293(November): 176–186. https://doi.org/10.1016/j.nucengdes.2015.07.031
9. Aakko-Saksa, Päivi T., Kati Lehtoranta, Niina Kuittinen, Anssi Järvinen, Jukka-Pekka Jalkanen, Kent Johnson, Heejung Jung, et al. 2023. "Reduction in Greenhouse Gas and Other Emissions from Ship Engines: Current Trends and Future Options." *Progress in Energy and Combustion Science* 94(January): 101055. https://doi.org/10.1016/j.pecs.2022.101055
10. Drosińska-Komor, Marta, Jerzy Głuch, Łukasz Breńkacz, Natalia Ziółkowska, Michał Piotrowicz, and Paweł Ziółkowski. 2025. "Advanced Genetic Algorithm-Based Signal Processing for Multi-Degradation Detection in Steam Turbines." *Mechanical Systems and Signal Processing* 224: 112166. https://doi.org/10.1016/j.ymssp.2024.112166
11. International Maritime Organization. n.d. "Autonomous Shipping." https://www.imo.org/en/MediaCentre/HotTopics/Pages/Autonomous-shipping.aspx.
12. Goerlandt, Floris. 2020. "Maritime Autonomous Surface Ships from a Risk Governance Perspective: Interpretation and Implications." *Safety Science* 128(August): 104758. https://doi.org/10.1016/j.ssci.2020.104758
13. Veitch, Erik, and Ole Andreas Alsos. 2022. "A Systematic Review of Human-AI Interaction in Autonomous Ship Systems." *Safety Science* 152(August): 105778. https://doi.org/10.1016/j.ssci.2022.105778
14. López, Manuel, and Virginia Antón. 2021. "Evolution of SBAS/EGNOS Enabled Devices in Maritime." *TransNav, International Journal on Marine Navigation and Safety Od Sea Transportation* 15(3): 541–549. https://doi.org/10.12716/1001.15.03.06
15. Lass, Christoph, and Ralf Ziebold. 2021. "Development of Precise Point Positioning Algorithm to Support Advanced Driver Assistant Functions for Inland Vessel Navigation." *TransNav, the International Journal on Marine Navigation and Safety of Sea Transportation* 15(4): 781–789. https://doi.org/10.12716/1001.15.04.09
16. "National Positioning, Navigation, and Timing (PNT) Architecture | US Department of Transportation." n.d. https://www.transportation.gov/pnt/national-positioning-navigation-and-timing-pnt-architecture.

17. Yang, Yuanxi. 2019. "Resilient PNT Concept Frame." *Journal of Geodesy and Geoinformation Science* 2(3): 1. https://doi.org/10.11947/j.JGGS.2019.0301
18. Department of Homeland Security. 2022. "Resilient Positioning, Navigation, and Timing (PNT) Conformance Framework Version 2.0." https://www.dhs.gov/sites/default/files/2022-05/22_0531_st_resilient_pnt_conformance_framework_v2.0.pdf (accessed 22 December 2025).
19. Dai, Zhen, Ralf Ziebold, and Evelin Engler. 2011. "The On-Board Maritime PNT Module – a Focus on Integrity Monitoring and Preliminary Results." *European Navigation Conference GNSS (ENC GNSS 2011)*, London, UK, vol. 28, 28 Nov–2 Dec 2011.
20. Szlapczynski, Rafal, Joanna Szlapczynska, and Roberto Vettor. 2023. "Ship Weather Routing Featuring W-MOEA/D and Uncertainty Handling." *Applied Soft Computing* 138: 110142. https://doi.org/10.1016/j.asoc.2023.110142
21. Szłapczyński, Rafał, Joanna Szłapczyńska, Mateusz Gil, Marcin Życzkowski, and Jakub Montewka. 2024. "Holistic Collision Avoidance Decision Support System for Watchkeeping Deck Officers." *Reliability Engineering & System Safety* 250: 110232. https://doi.org/10.1016/j.ress.2024.110232
22. Hem, Audun G., Edmund F. Brekke, Giorgio D. K. M. Kufoalor, and Ivan H. Kingman. 2024. "Autonomous Marine Collision Avoidance With Sensor Fusion of AIS and Radar." *IFAC-PapersOnLine, 15th IFAC Conference on Control Applications in Marine Systems, Robotics and Vehicles CAMS 2024*, Blacksburg, Virginia, vol. 58(20), 41–46. https://doi.org/10.1016/j.ifacol.2024.10.030
23. Kalman, Rudolph Emil. 1960. "A New Approach to Linear Filtering and Prediction Problems." *Journal of Basic Engineering* 82(1): 35–45
24. Filipowicz, Włodzimierz. 2014. "Fuzzy Evidence Reasoning and Navigational Position Fixing." *Recent Advances in Knowledge-Based Paradigms and Applications: Advances in Intelligent Systems and Computing*, vol. 234, Heidelberg, New York, London: Springer, 87–102.
25. Qiao, Shuanghu, Yunsheng Fan, Guofeng Wang, and Haoyan Zhang. 2023. "Multi-Sensor Data Fusion Method Based on Improved Evidence Theory." *Journal of Marine Science and Engineering* 11(6): 1142.
26. Dovis, Fabio. 2015. *GNSS Interference Threats and Countermeasures. Artech House GNSS Technology and Applications Series GNSS Interference, Threats, and Countermeasures.* Artech House.
27. Dai, Hai-fa, Hong-wei Bian, Rong-ying Wang, and Heng Ma. 2020. "An INS/GNSS Integrated Navigation in GNSS Denied Environment Using Recurrent Neural Network." *Defence Technology* 16(2): 334–340.
28. Saajasto, Mika, Maija Mäkelä, Fabricio S. Prol, M. Zahidul H. Bhuiyan, and Sanna Kaasalainen. 2023. "Convolutional Neural Network Based Approach for Estimating Ionospheric Delay from GNSS Observables." *2023 International Conference on Localization and GNSS (ICL-GNSS), 2023*, Castellón, Spain, 6–8 June 2023, 1–7. https://doi.org/10.1109/ICL-GNSS57829.2023.10148920
29. Belge, Egemen, Recep Fatih Cantekin, Berna Erol, Volkan Akgül, Seda Karadeniz Kartal, Rıfat Hacıoğlu, Sedar Kurtuluş Görmüş, Senol Hakan Kutoglu, and Kemal Leblebicioğlu. 2021. "Sensor Fusion Based on Integrated Navigation Data of Sea Surface Vehicle with Machine Learning Method." *2021 International Conference on INnovations in Intelligent SysTems and Applications (INISTA* 2021*)*, Kocaeli, Turkey, 25–27 Aug 2021, 1–6.
30. Deng, Liangkun, Xiang Zhang, Shiyong Tao, Ye Zhao, Keyi Wu, and Jie Liu. 2023. "A Spatiotemporal Graph Convolution-Based Model for Daily Runoff Prediction in a River Network with Non-Euclidean Topological Structure." *Stochastic Environmental Research and Risk Assessment* 37 (4): 1457–1478. https://doi.org/10.1007/s00477-022-02352-6
31. Zeng, Xi, Miao Gao, Anmin Zhang, Jixiang Zhu, Yingjun Hu, Pengxu Chen, Shuai Chen, Taoning Dong, Shenwen Zhang, and Peiru Shi. 2024. "Trajectories Prediction in Multi-Ship Encounters: Utilizing Graph Convolutional Neural Networks with GRU and Self-Attention Mechanism." *Computers and Electrical Engineering* 120(December): 109679. https://doi.org/10.1016/j.compeleceng.2024.109679
32. Squartini, S., A. Hussain, and F. Piazza. 2003. "Preprocessing Based Solution for the Vanishing Gradient Problem in Recurrent Neural Networks." *Proceedings of the 2003 International Symposium on Circuits and Systems* (ISCAS '03), Bangkok, Thailand, 25–28 May 2003, vol. 5 (May), V–V. https://doi.org/10.1109/ISCAS.2003.1206412

33. Chaal, Meriam, Xin Ren, Ahmad BahooToroody, Sunil Basnet, Victor Bolbot, Osiris A. Valdez Banda, and Pieter Van Gelder. 2023. "Research on Risk, Safety, and Reliability of Autonomous Ships: A Bibliometric Review." *Safety Science* 167: 106256. https://doi.org/10.1016/j.ssci.2023.106256
34. Munim, Ziaul Haque. 2019. "Autonomous Ships: A Review, Innovative Applications and Future Maritime Business Models." *Supply Chain Forum: An International Journal* 20(4): 266–279. https://doi.org/10.1080/16258312.2019.1631714
35. Kim, Tae-eun, Lokukaluge Prasad Perera, Magne-Petter Sollid, Bjørn-Morten Batalden, and Are Kristoffer Sydnes. 2022. "Safety Challenges Related to Autonomous Ships in Mixed Navigational Environments." *WMU Journal of Maritime Affairs* 21(2): 141–159. https://doi.org/10.1007/s13437-022-00277-z
36. Strode, Philip R. R., and Paul D. Groves. 2016. "GNSS Multipath Detection Using Three-Frequency Signal-to-Noise Measurements." *GPS Solutions* 20(3): 399–412. https://doi.org/10.1007/s10291-015-0449-1
37. Li, Binghao, Kai Zhao, and Xuesong Shen. 2020. "Dilution of Precision in Positioning Systems Using Both Angle of Arrival and Time of Arrival Measurements." *IEEE Access* 8: 192506–192516. https://doi.org/10.1109/ACCESS.2020.3033281
38. Kazim, Syed Ali. 2024. "*Enhancing Jamming Resilience in GNSS-Based Localization for Safety-Critical Land Transportation: Towards an Optimized Mitigation Approach Focusing User Requirements*." Phd thesis, Université Gustave Eiffel. https://univ-eiffel.hal.science/tel-04749087
39. Zamanzadeh Darban, Zahra, Geoffrey I. Webb, Shirui Pan, Charu Aggarwal, and Mahsa Salehi. 2024. "Deep Learning for Time Series Anomaly Detection: A Survey." *ACM Computing Surveys.* 57(1): 15:1–15:42. https://doi.org/10.1145/3691338
40. Wang, Senjie, and Zhengwei He. 2021. "A Prediction Model of Vessel Trajectory Based on Generative Adversarial Network." *Journal of Navigation* 74(5): 1161–1171. https://doi.org/10.1017/S0373463321000382
41. Zhang, Weibin, Weiyang Jiang, Qing Liu, and Weifeng Wang. 2023. "AIS Data Repair Model Based on Generative Adversarial Network." *Reliability Engineering & System Safety* 240 (December): 109572. https://doi.org/10.1016/j.ress.2023.109572
42. Liu, Hongdan, Yan Liu, and Zhengcong Zong. 2022. "Research on Ship Abnormal Behavior Detection Method Based on Graph Neural Network." *2022 IEEE International Conference on Mechatronics and Automation (ICMA 2022),* Guilin, Guangxi, China, 7–10 Aug 2022, 834–838. https://doi.org/10.1109/ICMA54519.2022.9856198
43. Xin, Rui, Jiale Pan, Fanlin Yang, Xiongfeng Yan, Bo Ai, and Qianran Zhang. 2024. "Graph Deep Learning Recognition of Port Ship Behavior Patterns from a Network Approach." *Ocean Engineering* 305(August): 117921. https://doi.org/10.1016/j.oceaneng.2024.117921

Index

Note: **Bold** page numbers refer to tables; *italic* page numbers refer to figures and page numbers followed by 'n' refer to notes.

For Product Safety Concerns and Information please contact our EU representative GPSR@taylorandfrancis.com Taylor & Francis Verlag GmbH, Kaufingerstraße 24, 80331 München, Germany

Batch number: 10004316

Printed by Printforce, the Netherlands

Strong Motion Instrumentation for Civil Engineering Structures [redacted]

JUN

NATO Science Series

A Series presenting the results of activities sponsored by the NATO Science Committee. The Series is published by IOS Press and Kluwer Academic Publishers, in conjunction with the NATO Scientific Affairs Division.

A. **Life Sciences**	IOS Press
B. **Physics**	Kluwer Academic Publishers
C. **Mathematical and Physical Sciences**	Kluwer Academic Publishers
D. **Behavioural and Social Sciences**	Kluwer Academic Publishers
E. **Applied Sciences**	Kluwer Academic Publishers
F. **Computer and Systems Sciences**	IOS Press
1. **Disarmament Technologies**	Kluwer Academic Publishers
2. **Environmental Security**	Kluwer Academic Publishers
3. **High Technology**	Kluwer Academic Publishers
4. **Science and Technology Policy**	IOS Press
5. **Computer Networking**	IOS Press

NATO-PCO-DATA BASE

The NATO Science Series continues the series of books published formerly in the NATO ASI Series. An electronic index to the NATO ASI Series provides full bibliographical references (with keywords and/or abstracts) to more than 50000 contributions from international scientists published in all sections of the NATO ASI Series.
Access to the NATO-PCO-DATA BASE is possible via CD-ROM "NATO-PCO-DATA BASE" with user-friendly retrieval software in English, French and German (WTV GmbH and DATAWARE Technologies Inc. 1989).

The CD-ROM of the NATO ASI Series can be ordered from: PCO, Overijse, Belgium

Series E: Applied Sciences – Vol. 373